Integrated Assessment for Water Framework Directive Implementation

Water Framework Directive Series

Integrated Assessment for Water Framework Directive Implementation

Data, Economic and Human Dimension

Volume 2

Edited by
Peter A. Vanrolleghem

Publishing
London • New York

Published by **IWA Publishing**
Alliance House
12 Caxton Street
London SW1H 0QS, UK
Telephone: +44 (0)20 7654 5500
Fax: +44 (0)20 7654 5555
Email: publications@iwap.co.uk
Web: www.iwapublishing.com

First published 2010

Typeset in India by OKS Prepress Services.

British Library Cataloguing in Publication Data
A CIP catalogue record for this book is available from the British Library

Library of Congress Cataloging-in-Publication Data
A catalog record for this book is available from the Library of Congress

ISBN 10: 1843393263
ISBN 13: 9781843393269

Contents

Authors for this volume xiii
Preface xvii

***Guidance Report II.1* Improving integrated assessment in model-supported river basin management** **1**

Preface 1
Keywords 2
1. Integrated Assessment – Issues Addressed 2
2. Supporting River Basin Management with Computer-based Modelling Tools 4
   2.1. The role of modelling in water resources management: expectations and reality 5
   2.2. Modelling and decision making in water resources management: (dis-)entangling two processes 6
   2.3. Confidence in modelling tools: a matter of uncertainty? 7

2.4. Looking at the interface: communication and user-friendliness 8
2.5. Participatory approaches: acknowledging the human dimension in modelling and management 8

3. Recommendations for Improving the Structure of the Science-policy Interface 9
3.1. Communication 9
3.2. Incentives for interaction: recommendations for funding 10
3.3. Managing of expectations-reflections on the role of research 12

4. Content-related Issues for a Future Research Agenda 12
4.1. Issues to be addressed by models to support the WFD 13
4.2. Economic aspects of the WFD 13
4.3. Data availability, accessibility and use in modelling 13

5. Summary and Conclusion 14

6. References 15

7. Annex: Research Directions Identified in Harmoni-CA/WP5 Workshops 18

Participatory Research Processes and Research Directions from 1st Methods Workshop 18

Recommendations of 2nd Harmoni-CA/WP5 Policy Workshop on Models Supporting the Interaction of Agricultural and Water Management 21

Summary and Conclusions from 4th Harmoni-CA/WP5 Policy Workshop: Elements of a Future Research Agenda 23

***Guidance Report II.2* Data availability and accessibility in view of model-based water management 27**

Preface 27

Keywords 28

Executive Summary 28

1. Data for Modelling – What are the Problems? 31
1.1. Why is modelling an important tool in water resources management 31
1.2. Objectives and target audience of this document 32
1.3. Water Framework Directive – requirements 32

1.4. Do we have enough data for modelling? 34
1.5. Are the existing data accessible for modelling? 35

2. Data Requirements – Methodological Framework 35
2.1. Modelling categories 35
2.2. Data requirements for the WFD implementation 39
2.3. Data requirements for modelling 41
2.4. How and when in the modelling process is the necessary amount of data assessed? 43
2.5. The relation between data availability and model performance 46
2.6. Conclusions on methodological considerations regarding data requirements 47

3. Data Availability 48
3.1. Methodological considerations 48
3.2. Information from WFD Article 5 reports (Basis Analysis) 49
3.3. HarmoniRiB River Basin Network 50
3.4. Discussion and conclusions regarding data availability in river basins 57

4. Data Accessibility 59
4.1. Accessibility for water resources management 59
4.2. Accessibility for research purposes 61

5. Data Quality – Uncertainty 62

6. Conclusions and Recommendations 64
6.1. Data availability 64
6.2. Data accessibility 65
6.3. Data quality – uncertainty 66
6.4. Data for research purposes 67

7. References 68

8. Appendix – HarmoniRiB River Basins 71

***Guidance Report II.3* Good practice in joint use of monitoring and modelling 81**

Preface 81

Keywords 82

1. Executive Summary 82

2. Introduction 84

2.1. Why joint use? ... 85
2.2. What are the problems? ... 86
2.3. Objectives and target audience of this document ... 88

3. Monitoring and Modelling within the WFD ... 89
3.1. WFD monitoring programmes ... 89
3.2. Modelling categories ... 92
3.3. Balancing precision, data and model concept ... 96

4. Joint Use of Monitoring and Modelling ... 97
4.1. Operational modelling support for monitoring ... 99
4.1.1. Quality assurance ... 99
4.1.2. Interpolation and extrapolation in time and space ... 100
4.1.3. Conceptual model ... 102
4.1.4. Assess effects of anthropogenic activities ... 103
4.1.5. Design of monitoring programme ... 105
4.2. Perspectives for future modelling support for monitoring ... 107
4.2.1. Model development ... 107
4.2.2. Uncertainty reduction in models ... 108
4.2.3. Flood forecasting ... 108

5. Discussion and Conclusions ... 109

6. References ... 111

7. Appendix – Case studies ... 113
7.1. The national water resources model for Denmark (DK-model) ... 114
7.1.1. DK-model structure and purpose ... 114
7.1.2. DK-model and monitoring ... 114
7.1.3. References ... 116
7.2. Matsalu River Catchment ... 116
7.2.1. Modelling area and purpose ... 116
7.2.2. Matsalu model ... 117
7.2.3. Matsalu model and monitoring ... 119
7.2.4. References ... 120
7.3. National Hydrological Forecasting System of Hungary (NHFS) ... 120
7.3.1. Structure and operation of the NHFS ... 120
7.3.2. NHFS and monitoring ... 121
7.3.3. References ... 126
7.4. Groundwater model in monitoring network design (Monitoring Design) ... 127

8. Appendix – Common Flow Chart 129
8.1. Description of steps in flowchart: 130
8.2. Additional modelling support 133
8.2.1. Quality assurance of data 133
8.2.2. Interpolation/extrapolation 134
8.2.3. Optimisation of monitoring programmes 134
8.3. Data 134
8.3.1. References 134

Guidance Report II.4 **Understanding the role of perception and valuation in the development and use of models for water management 135**

Abstract 135
Keywords 135
1. Introduction 136
1.1. Models, modelling and the use of models in water management 138
1.1.1. Types of models 140
1.1.2. Use of models 141
1.2. Actors in model-based water management 143
1.2.1. Archetypal actors in a modelling context 143
1.2.2. Some typical interactions 146
1.3. The human dimension in model-based water management 148
1.3.1. Model development 148
1.3.2. Model use 149
2. Human Dimension in the Development of Models 150
2.1. Perception, framing and model development 151
2.1.1. Frames, beliefs and values 151
2.1.2. Re-humanizing our representation of the modelling process 153
2.1.3. Case study I – Examining influences on integrated assessment model development 157
2.1.4. Managing differences in perception and frame 160
2.1.5. Case study II – How frames influence model bounding decisions 165
2.2. Incorporating human needs, wants, values and behaviours into models 166
2.2.1. Representing the human dimension in a model 166

2.2.2. Scenario modelling: how perceptions influence scenario modelling and use 172
2.2.3. Case study III – The use of scenarios in the MANTRA-East Project 174
2.2.4. Participatory modelling – What kind of input you are after? 175
2.2.5. Case study IV – Participatory modelling in South Africa 177
3. Human Dimensions in the use of Models 178
3.1. Introduction 178
3.2. What makes a model useful? 182
3.3. Models as tools to support action 188
3.4. Data, information and knowledge 188
3.5. The role of management and managers: how are models interpreted? 190
3.6. The strategic use of models in organizations 191
3.7. Political use of models 193
3.8. Case study V – Participatory stakeholder analysis for water management in Egypt 194
4. Recommendations 197
4.1. Guidance for model commissioners 197
4.2. Guidance for model developers 198
4.3. Guidance for model users 199
5. Conclusion 199
6. References 199

***Guidance Report II.5* Economic methods, models and instruments for the Water Framework Directive 207**

Abstract 207
Keywords 207
1. Introduction 207
2. The Economic Analysis in the WFD 209
3. Key Issues in the Economic Analysis in the WFD 212
3.1. Integrated water-economic river basin information systems and indicators 213
3.2. Selection of cost-effective programs of measures to reach the WFD objectives 216

3.3. Definition and assessment of disproportionate costs to reach the WFD objectives 221
3.4. Cost recovery of water services 224
3.5. Environmental and resource costs 229
3.6. Design, application and evaluation of economic instruments 234

4. Economic Methods, Models and Instruments 239
4.1. Introduction 239
4.2. Cost-effectiveness analysis 240
4.3. Cost-benefit analysis 244
4.4. Non-market valuation methods 247
4.5. Benefits transfer 251
4.6. Multi-criteria analysis 255
4.7. Integrated river basin models 258
4.8. Economic instruments 266

5. Summary and Conclusions 270

6. References 271

7. Annex: Review of Selected Integrated Hydro-economic Models 276

Authors for this volume

David N. Barton
Norwegian Institute for Nature Research (NINA)
Oslo, Norway
E-mail: david.barton@nina.no

Ilke Borowski
seeconsult GmbH
Osnabrück, Germany
E-mail: iborowski@seeconsult.org

Pieter Bots
Faculty of Technology, Policy and Management
Delft University of Technology
The Netherlands
E-mail: p.w.g.bots@tudelft.nl

Roy Brouwer
Department of Environmental Economics Institute for Environmental Studies
VU University Amsterdam
The Netherlands
E-mail: roy.brouwer@ivm.vu.nl

Lisbeth Flindt Jørgensen
Department of Hydrology
Geological Survey of Denmark and Greenland
Copenhagen, Denmark
E-mail: lfj@geus.dk

Geoffrey Gooch
Center for Water law, Policy and Science,
University of Dundee, Scotland &
Linköping University, Sweden
E-mail: geoffrey.gooch@liu.se

Anker Lajer Højberg
Department of Hydrology
Geological Survey of Denmark and Greenland
Copenhagen, Denmark
E-mail: alh@geus.dk

Brian S. McIntosh
Cranfield University
United Kingdom
E-mail: B.McIntosh@cranfield.ac.uk

Frans Oosterhuis
Department of Environmental Economics Institute for Environmental Studies
VU University Amsterdam
The Netherlands
E-mail: frans.oosterhuis@ivm.vu.nl

Claudia Pahl-Wostl
Institute of Environmental Systems Research
University of Osnabrück, Germany
E-mail: pahl@usf.uni-osnabrueck.de

Jens Christian Refsgaard
Department of Hydrology
Geological Survey of Denmark and Greenland
Copenhagen, Denmark
E-mail: jcr@geus.dk

Caroline van Bers
seeconsult GmbH
Osnabrueck, Germany
E-mail: cvanbers@seeconsult.org

Frans van Geer
TNO Built Environment and Geosciences
Geological survey of the Netherlands
TA Utrecht, The Netherlands
E-mail: Frans.vangeer@tno.nl

István Zsuffa
VITUKI Environmental and Water Management Research Institute Non-profit Ltd.
Budapest, Hungary
E-mail: istvan.zsuffa@vituki.hu

Preface

Water resources planning and management and the development of appropriate policies require methodologies and tools that are able to support systematic, integrative and multidisciplinary assessments at various scales. It also requires the quantification of various uncertainties in both data and models, and the incorporation of stakeholders participation and institutional mechanisms into the various tools and risk assessment methodologies, to help decision makers understand and evaluate alternative measures and decisions. This requirement has been explicitly recognized in the context of the European Union Water Framework Directive (WFD). Designing the Programme of Measures is one of the key elements in building up the river basin management plan as defined by the WFD. The planning process for producing the Programme of Measures consists of several consecutive steps, starting from setting the objectives to the analysis of alternative ways of reaching them. It becomes evident that this is a challenging task which needs support from appropriate information systems and modelling tools that are able to cope with the complexity of the water system and planning process. This is particularly important in the context of current discussions on climate change impacts on the water sectors and actions needed

 Integrated Assessment for Water Framework Directive Implementation Water Framework Directive Series Volume 2. Edited by Peter Vanrolleghem. ISBN: 9781843393269. Published by IWA Publishing, London, UK.

to adapt to those changes. Since changes in the availability of water resources and occurrence of extreme events will affect not only the water supply and sanitation services but also other economic sectors, like agriculture, forestry, tourism, inland navigation, hydropower, health, etc. it is important to prepare a framework which allows policy makers to make integrated assessments.

To respond to this challenge, the European Commission has initiated in the context of the 5th EU Framework Programme for Research and Technological Development (FP5) a large CatchMod initiative. Its objective was to develop and harmonize integrated modelling tools that could be used by water authorities for planning and managing water resources in an integrated way at the scale of river basins. This very ambitious initiative with long-term objectives has been implemented trough a dedicated cluster constituted by a core group of several research projects and coordinated by a concerted action, the Harmoni-CA project. CatchMod and Harmoni-CA addressed several key issues in modelling research, including model concepts and regimes, scale issues and data availability, understanding stakeholders participation, participatory approaches and institutional mechanisms and their incorporation in the various biophysical models and risk assessment methodologies to help decision-makers understand and evaluate alternative approaches, etc. Science-policy integration is another important challenge that has been addressed by the CatchMod initiative and the Harmoni-CA project.

In many instances, the lack of communication and of clear coordination mechanism hinders the research outputs from being used or simply known by policy makers, and hinders policy research needs from being communicated to the scientific communities in a timely fashion. By initiating and facilitating the discussion between the Common Implementation Strategy Coordination Group of the WFD and the various researchers, Harmoni-CA helped to foster synergies between researchers, policy makers and water managers on different levels to achieve a common understanding of the various WFD implementation issues. It also helped consolidate knowledge generated by several EC projects (of the CatchMod cluster and beyond) on various cross-cutting and complementary research issues, with a view of providing technical support for the implementation of the WFD.

This publication is one of the concrete outcomes of the Harmoni-CA concerted action as part of a 4-volume series of Guidance Reports that guide water professionals through the implementation process of the Water Framework Directive, with a focus on the use of ICT-tools (and in particular modelling). They are supposed to be complementary to the Guidance Documents produced by the EU Directorate General for Environment. An initial volume "**Water Framework Directive: Model supported Implementation – A Water Manager's Guide**", edited by Fred Hattermann and Zbigniew W Kundzewicz, provides a practical guidance to water managers and other interested stakeholders on the model-supported implementation of the WFD. The three other volumes bring more detailed guidance regarding a range of aspects of such WFD-implementation. This particular volume 2 focuses on the "**Integrated Assessment for WFD implementation: Data, economic and human dimension**", containing five Guidance Reports: (i) Improving integrated assessment, (ii) Data availability and accessibility, (iii) Joint use of monitoring and modelling, (iv) The role of perception and valuation in the development and use of models and (v) Economic methods, models and instruments. The two other volumes cover "**Modelling aspects of WFD implementation**" in four additional Guidance Reports and "**Decision support for WFD implementation**" with two final Guidance Reports. These Guidance Reports were written as part of the work conducted within the different CatchMod cluster projects. Therefore, some of these reports were already finalized in 2005–2006 and report the status at the time. Evidently, research and practice have moved forward, but it is expected that the review and synthesis nature of the Guidance Reports makes that the information they contain is still very actual.

These four volumes provide a good example of disseminating research results of policy relevance. We hope that this publication will reach a broad audience, both science and practice oriented, thus helping to further strengthen the links between scientific research, policy and decision making.

Dr. Panagiotis Balabanis
Principal Scientific Officer
Directorate General for Research
Commission of European Communities

Prof. Peter A. Vanrolleghem, Editor
Current address: modelEAU
Université Laval
Québec (QC) Canada

Guidance Report II.1

Improving integrated assessment in model-supported river basin management[1]

Ilke Borowski, Caroline van Bers and Claudia Pahl-Wostl

Preface

This Guidance Report was originally produced within the framework of Harmoni-CA – Harmonised Modelling Tools for integrated River Basin Management (EVK1- CT-2002-2003, www.harmoni-ca.info). Harmoni-CA is a concerted action supported by the European Commission under the Fifth Framework Programme and contributing to the implementation of the Key Action "Sustainable Management and Quality of Water (EKVI-2001-00192). The purpose of Harmoni-CA was to facilitate the dialogue and to help bridging the gap between research and policy. This was achieved by synthesizing available knowledge produced by the various Research – Technology and Development projects in the area of catchment modelling (CatchMod). The project also facilitated the development of methodologies and tools to support the use of Information, communication and technology (ICT) tools in implementing the Water Framework Directive (WFD).

Harmoni-CA work package 5 (WP5) focused on "Integrated Assessment and the science-policy interface". The specific task has been to develop and strengthen the science-policy interface across sectors and spatial boundaries. Part of this task has involved the creation of opportunities for water managers to learn about related research activities and for researchers to learn about water managers' needs. A central aspect of WP5 was to stimulate both groups to reflect more on their work. Thus a total of 9 workshops were organized (including four policy workshops, three method workshops, a synthesis workshop, a joint workshop with the International Water Association and The Integrated Assessment Society) on various methodological and policy-orientated issues in order to increase the understanding of integrating the human dimension into river basin management (RBM) and in the implementation of the WFD. The choice of themes addressed was the outcome of consultation with the WP5 task force consisting of stakeholders and experts in the field of RBM, namely two water managers and three researchers from five different countries.

This report synthesizes the results of the European Concerted Action Harmoni-CA – specifically those related to insights on integrated assessment and the science-policy

[1]An earlier version of this document was developed as part of Harmoni-CA/WP5, deliverable D 5.4d

 Integrated Assessment for Water Framework Directive Implementation Water Framework Directive Series Volume 2. Edited by Peter Vanrolleghem. ISBN: 9781843393269. Published by IWA Publishing, London, UK.

interface. It provides recommendations concerning the practice of integrated assessment and priorities in the development of new RTD activities.

Keywords
Communication, Human Dimension, Integrated Assessment, Participatory Processes, Science-Policy Interface

1. INTEGRATED ASSESSMENT – ISSUES ADDRESSED

"*Integrated Assessment (IA) can be defined as the interdisciplinary process of integrating knowledge from various disciplines and stakeholder groups in order to evaluate a problem situation from a variety of perspectives and provide support for its solution. IA supports learning and decision processes and helps to identify desirable and possible options for addressing the problem. It therefore builds on two major methodological pillars: approaches to integrating knowledge about a problem domain, and understanding policy and decision making processes*" (The Integrated Assessment Society, www.tias-web.info).

Being located at the boundaries is intriguing but also gives rise to dangers. Integrated assessment faces the challenge of earning credibility in both the scientific and the policy communities and is at risk of not being respected by either. Hence, a forum for the exchange of experience is urgently needed.

IA in the context of integrated water resources management (IWRM) requires the integration of knowledge across scales, across boundaries, between different components of the water system, across various scientific disciplines, and/or policy fields (Gilbert *et al.*, 1998). Also, in the context of river basin management as prescribed by the European Water Framework Directive (WFD; (2000/60/EC), the challenge of integrating knowledge has been identified as one of the main challenges – especially given the tight timeline demanded by the directive in order to achieve a satisfactory ecological state (European Commission 2001, 2004; de Rooy *et al.*, 2005). Typically, issues of integrated assessment in water management, however, address not only specific water-related topics (e.g., eutrophication, water deficits, floods, agriculture, or urban areas) but they also include umbrella issues such as complexity, integration across different scales and/or uncertainty that tend to dominate the discussion (Gilbert *et al.*, 1998).

Harmoni-CA has addressed issues related to integrated assessment and the science-policy interface in the context of IWRM from various angles. Two approaches were applied here: meetings were organized for water managers and researchers to foster exchange and communication; and research insights were synthesized in several documents to provide guidance on modelling-supported water management. For example, a water manager's guide for model supported implementation of the WFD has been developed to "*contribute to improvement of the understanding of modelling aspects to facilitate the planning process for the implementation of the Directive*" (Hattermann and Kundzewicz, 2009). Also, the generation and use of data as one of the very central aspects of successful model support in RBM have been addressed in the context of the joint use of modelling and monitoring. As Højberg *et al.* (2010) puts it: "*The overall objective of the Water Framework Directive is to achieve a good quantitative and qualitative status of all waters by 2015*".

The WFD brings two important challenges: 1) water must be managed in an integrated way, thus including all uses of water and all domains, such as surface- and groundwater, and 2) the qualitative status is related not only to the chemical substances measured in the water, but should also be evaluated based on the ecological status. To fulfil these requirements development of new tools and models is a necessity. To avoid a future, parallel development of monitoring and modelling praxis in the field of integrated and ecological assessments, it is of vital importance that monitoring and modelling are realised to be inter-linked activities and not independent disciplines. Two Guidance Reports have been published exploring the need and options for joint use of monitoring and modelling and raising awareness of the accessibility of data issue (Højberg *et al.*, 2010; Refsgaard *et al.*, 2010b). They link to existing efforts for improving the accessibility of water related tools and information such as WISE (Water Information System for Europe; www. http://www.water.europa.eu/) and INSPIRE (Infrastructure for Spatial Information in Europe http://sdi.jrc.it/inspire). Both reports include recommendations for improving data availability, accessibility, and use for monitoring and modelling. In addition, other aspects of technical catchment modelling of IA were addressed by Harmoni-CA such as uncertainty analysis, sensitivity analysis, model calibration, and quality assurance in modelling, as reported in Volume 1 of this series (Vanrolleghem, 2010). Also, specific research results from the catchment modelling projects can be found in the web portal WISE-RTD (http://wise-rtd.info/) which was also developed by Harmoni-CA.

Perhaps most importantly, Harmoni-CA/WP5 went far beyond the technical aspects of modelling supported water management in addressing the science-policy interface. WP5 focused strongly on the challenges of integration of different fields and disciplines into water management for the purposes of fulfilling the WFD, such as economy (Heinz *et al.*, 2007; Brouwer *et al.*, 2010), agriculture (Bazzani, 2010; Borowski and Heeb, 2005), and participatory aspects (Hare, 2003). Furthermore, the human dimensions of water management and modelling in terms perceptions and values have been a focus (Bots *et al.*, 2010). Many valuable insights emerged on the science-policy interface and how to improve it (Borowski and Hare, 2007a; Brugnach *et al.*, 2007; Giupponi *et al.*, 2010; Horzlitz, 2007; Rotter and Rasche, 2007). By setting up a series of interactive and output oriented workshops, WP5 contributed both to capacity building for bridging the gap between science and policy; and to identifying further research directions and other recommendations for narrowing the gap.

This Guidance Report synthesizes the insights from Harmoni-CA/WP5 workshops and reports into recommendations for further improving the science-policy interface. The report contributes thus to a strengthening of the uptake of new research results in management and promotion of integrated water resources management. The report does not provide a complete review of literature related to the various issues presented here. References are made primarily to documents that were developed in the Harmoni-CA context. The intention, in particular, is to provide recommendations for better uptake of model-based tools that were developed and tested in various research projects. That means that our main concern is the use of research results and we consider this in the context of modelling and more generally ICT tools.

To enable a better understanding of the background of the recommendations, the report first summarizes briefly those issues that persisted throughout Harmoni-CA discussions. Section 3 then provides recommendations for improving the structure of the science-policy interface while section 4 presents the research directions identified. Section 5 concludes the discussion by briefly reflecting on the recommendations.

2. SUPPORTING RIVER BASIN MANAGEMENT WITH COMPUTER-BASED MODELLING TOOLS

The discussions during the eight workshops and five river basin meetings covered a range of issues related to how models can support the implementation of the WFD. However, several of these issues prevailed in the various discussions that took place. In this section, we have condensed them into five issues which we introduce in this section to provide the reader with more background. These are the role of models, the modelling- and the decision making process, confidence levels, communication and user-friendliness, and finally, the human dimension (see also Borowski and Hare, 2007a).

2.1. The role of modelling in water resources management: expectations and reality

In the context of this Guidance Report, the use of models in water management in general has to be distinguished from the use of research-derived models. Almost all model software used in daily practice are standard tools (and not only empirical tools, but also complex model codes like MODFLOW, FEFLOW, SOBEK and MIKE SHE) that were originally developed in a research context. By research-derived modelling tools, we mean those tools that are currently developed, tested, and/ or applied within the frame of research projects. Harmoni-CA acknowledges that model-based tools play an important role in supporting water resources management (Hattermann and Kundzewicz, 2009). Senior modellers at the first Harmoni-CA/WP5 Modelling Workshop even saw the process of modelling itself as being central to water resource management planning. Hence, there was agreement that participatory modelling is a central element for participatory management (Hare, 2004).

However, in day-to-day practice, water managers have considered models as one of many sources of information that contribute to a final decision (Hare, 2005). This seems to be strongly linked to the liability and responsibility associated with the decisions taken by water managers that require them to consider as many sources of knowledge as possible. In some countries such as Denmark or Germany, the application of models for decision making purposes is standard practice. In other countries, the application of modelling tools is in its infancy. Often, models are developed for addressing hydrological issues such as flood migration in which much data is required and the users have to be experts. There is an evident lack of socio-economic models and representation of human dimensions such as stakeholder interests (Hare, 2005). However, the expectations of water managers were often to improve the role of models in order to deliver specific and reliable answers to specific questions (Borowski and Hare, 2007b; Hare, 2003; Interwies and Borowski, 2007).

Reflecting these insights, Højberg *et al.* (2010), state that "*a shift in the perception of models is needed, where models currently often are considered useful only to provide answers to specific questions. This view must be replaced by recognition of models as being valuable tools that should be used continuously and interactively along with monitoring data, a tool that serves many purposes from the design phase of monitoring programmes to the interpretation and quality assurance of the monitoring data*". Model-based tools have played a smaller role in participatory approaches. At the same time, water managers have a higher expectation that models developed by researchers can contribute to improved understanding between various stakeholders. Yet, the problem for participatory use of models is the same as that of participatory management itself: how do you get people to start using the models and then keep them using them? The solution is not technological. Rather it is sociological/political: there is a need for a paradigm shift in the way stakeholders and/or the public are engaged in management processes. Stakeholder acceptance of model-based management practices is also often limited since new models are difficult to explain and ultimately to sell to lay people (Borowski and Hare, 2007b; Hare, 2005).

During the workshops, support for water management by research-derived model-based tools was more often identified in the process of model building and use. The outcomes (numbers, graphics) or the model itself were less important. This means that it is not the data or the maps as outputs of a model that provide the most important insights as such, but the process of learning about the system which is being modelled that provides more support to model-users. Stakeholder involvement in research projects is obviously an effective means for improving the impact of research.

Improving stakeholder involvement with participatory modelling in decision-making processes requires that model developers become more aware of the needs of the stakeholders as policy makers and their operating constraints. In this way, models can be developed that target policy relevant issues by integrating specific norms or indicators. For example, the distribution of hazardous substances modelled needs to cover those substances which have been identified as a pressure in the modelled basin and requiring monitoring (e.g., as part of list of priority hazardous substances according to Annex X of the WFD). Such models could help water managers to create the processes and institutions which help to identify the interaction of the various interests and uses. This would be a first step for establishing a learning process among stakeholders in order to develop socially, economically and ecologically sustainable solutions in RBM.

2.2. Modelling and decision making in water resources management: (dis-)entangling two processes

One reason why models are often not applied in water management lies in the different processes for modelling and decision making. In the context of European-funded research, modelling processes take place within the scope and time frame of research projects, (i.e., 3–5 years). A complete modelling process including the non-linear process of problem definition, software choice and implementation, validation, and implementation in various locations or case studies, commonly requires about ten years. In addition, water management processes as conducted during the implementation of the WFD, demand river basin plans or a programme of measures within a demand river basin plans or a programme of measures within a legally-prescribed time frame. Often, water authorities are restricted by budgetary limitations. In general, financial resources and thus personnel are as limited as time. Thus, newly developed tools in research projects cannot be expected to be applied directly in decision making processes. Application in practical water management is therefore more promising if existing tools are further improved and adapted to the specific basin characteristics. Unfortunately, such case-study based research has received less scientific recognition.

2.3. Confidence in modelling tools: a matter of uncertainty?

Although water managers understand the potential value of using models, they have also been rather critical in the course of the project. They seemed to lack confidence in models- not only research-derived models, but all models that had not yet been tested and validated (Borowski and Hare, 2007a). Researchers acknowledge that users may lack confidence in the models they develop. To establish confidence in a model, sound validation is needed but other factors that play a role include the "*mathematics of the model; the reliability of the model in terms of predicting processes in nature; use & familiarity; simplicity and complexity. It includes convincing people of the merits of a model and the trust that comes with experience of a model and with modelling. Water managers ask for transparent models with their limitations clear and communicated*" (Borowski and Hare, 2007a).

However, for lack of alternatives in measuring the reliability of a model, uncertainty analysis is in general called for to establish a level of trust in a modelling tool. Quantifying uncertainty remains a challenge, especially with respect to data collection. One reason may be the lack of available methodologies, or insufficient time to identify and assess available knowledge on these methodologies (van Bers *et al.*, 2007). In approaching uncertainty, scientists have to be more explicit about probabilities, and policy makers have to be more explicit about required confidence levels for decision-making. In such a process, uncertainty analysis can only address uncertainties to a certain extent. Building confidence and trust within decision making obviously goes far beyond this. More insights on the challenges of uncertainty analysis and ways to address these challenges can be found in Brugnach *et al.* (2007) or Refsgaard *et al.* (2010a). However, if transparency in methods and scientific results is not achieved, models will continue to be perceived as an obstacle to bridging the science-policy gap (van Bers *et al.*, 2007).

2.4. Looking at the interface: communication and user-friendliness

In examining the interface (i.e., the interaction) between the research community and water management community, three needs dominated the Harmoni-CA discussion:

(1) an ergonomically-sound interface for the computer-based modelling tools that helps users to understand the tools and make use of the results,
(2) accessible information on validated tools,
(3) skills in social learning, (i.e., communication of scope and limits of the tool leading to shared understanding and enabling improved joint development of the tool and eventually also of the issue modelled).

With respect to the first two needs, Horlitz (2007) has pointed out that for end-users access to information concerning management options plays a vital role. Increasingly, large knowledge bases along with a wider array of models are being developed for water management that address the 'what if' questions. The extent to which this information and knowledge as well as the models and tools themselves can be made available and useable for decision making and for participatory processes in general, appears to be limited by restricted access to appropriate hardware and the internet and a lack of suitable interfaces and internet platforms. Even if an interface is adequate, it must often be complemented with face-to-face meetings. During a participatory management process, a model needs to be facilitated by someone who is knowledgeable in the use of the model (Horlitz, 2007).

The need for social learning was often mentioned in the context of participatory approaches and will be discussed in the next section.

2.5. Participatory approaches: acknowledging the human dimension in modelling and management

The individuals involved in the Harmoni-CA project often stood "on different sides of the model" and needed to take the step of integrating their collective knowledge and experiences: "*Our experience is that statisticians, policy-makers, social scientists, natural scientists and all others involved in the fascinating business of modelling have much to give and much to learn from each other. And after all, all our models are created and used by humans, so the human dimension cannot be isolated from all the others*" (Bots *et al.*, 2010).

Participatory approaches become almost mandatory especially when faced with the complexity of reconciling different interests in the same river basin, to ensure the adaptiveness of the basin, that is to say, the potential to respond to new insights or correct earlier decisions (Pahl-Wostl, 2002; Pahl-Wostl, 2007a).

The improvement in the application of tools in (participatory) decision making as well as the improved integration of results from social and natural science, can benefit from shared learning which goes deeper than simply listening to each other's presentations. "*Those learning processes required at the individual level include the ability to listen, reflect and also ensure understanding of new issues - an ability increasingly assigned to ambassadors or intermediaries. (...) it became clear that the improved integration of social sciences in water modelling would likely also benefit the application of models in the decision making process*" (van Bers *et al.*, 2007).

Stakeholder participation in research projects or in water management also implies the need for a new approach to dealing with uncertainty and a new perspective on policy processes (see section 2.3). The scientific community can learn a great deal here from the rich experience of the Integrated Assessment community, in particular in the context of climate change.

3. RECOMMENDATIONS FOR IMPROVING THE STRUCTURE OF THE SCIENCE-POLICY INTERFACE

3.1. Communication

When attempting to improve communication between different groups (e.g., water managers, researchers, funding bodies and water policy makers), the basis for fruitful exchange is improved understanding of each other and learning jointly. Communication between research and water management includes the dissemination of information not only by technical means but also between and among people.

Mechanisms for information dissemination, such as the WISE-RTD web portal (wise-rtd.info) that provides among other types of information, ICT tools that support the implementation of the European Water Framework Directive, need to be freely accessible (both of upload and download of information) and then strongly promoted. Such an instrument may not be sufficient in providing all available information, but it can serve stakeholders with a range of perspectives and interests if it allow tailored access points for the different target audiences. An array of data bases, web portals, and/or data access points is necessary to serve the needs of diverse stakeholders. This means that centralization of information can only be successful to a certain point. Thus, information sources and web portals should instead be interlinked with each other than being centralised into one source or portal.

In addition to this technical consideration, successful communication between research and water managers occurs when members of the groups engage directly. This is to say that they reflect jointly and exchange perspectives on their problems. Often, however, the necessary communication skills are lacking and hamper learning processes.

Therefore, Harmoni-CA strongly recommends that training in communication is promoted and provided for both researches and water managers. Water managers need to pose specific questions in terms of the problems to be solved. Researchers need to reflect more on these questions and adapt their research towards responding to them or they need to be more transparent on the boundaries of their research with respect to the questions posed. This training should include not only rhetorical and presentation skills. Social learning, including listening, critical reflection and evaluation of joint work, is especially helpful for successful cooperation. Such training would contribute to the basics of Integrated Assessment, namely the exchange and harmonization of the various concepts and cause-effect models that can be found in the different disciplines (Newell *et al.*, 2005; Pahl-Wostl, 2007b).

In addition, network activities in the various EU member states should be encouraged. As is the case with mechanisms for information dissemination, social networks establish themselves around shared themes or interests. They cannot grow endlessly. Thus, for different target groups, different networks need to be promoted. For example, national networks in water management could start from "focal points" where experience from management and (regional) research could be brought together and facilitate exchange and national harmonisation where appropriate.

3.2. Incentives for interaction: recommendations for funding

Developing a model from scratch is expensive and requires much expertise and time. For water authorities, it is crucial to minimize the risk that the model development process takes longer to complete than anticipated, or that an unusable or unreliable tool is delivered (Bots *et al.*, 2010). When working exclusively with research institutes one must be aware that such risks are higher and amplified if the time frame for development is too restricted. Often the continuity of personnel and expertise is affected by the project-based employment strategies of many universities.

If the aim of European-based research is to further promote the development of models for water resources management, two issues need to be addressed: the time allocated to research for water management and water management processes themselves; and the incentives to encourage interaction between researchers and water managers.

With respect to the first issue, research projects are generally 3–5 years in length and need to follow the projects' schedule to secure funding that is allocated annually. Case-study oriented research which is in general required for water management involving highly specific management-related questions usually requires a shorter timeline. Another disadvantage for researchers is that such studies are less likely to reveal new scientific insights and therefore result in fewer high-ranking publications. The latter, however, provide a key indicator for assessing scientific expertise of researchers when applying for project funding. If scientists need to focus first on approaches that are compatible with the given time line and therefore easier to implement, this might suggest that changes are needed in the general incentive structure for scientists which asks strongly for high-level scientific publications and less for fast implementable tools. Strengthening of case study approaches in research could in addition be improved through pilot implementations of new instruments and the replication of results, demonstrating examples of "good practice" in various water management basins.

Thus, an alignment of funding cycles and policy processes may not be sufficient since innovative research is usually stronger if given more time. Involvement of water authorities should be (financially) encouraged in earlier phases of proposal planning. Alternatively, other funding forms allowing continuous support of water management processes by research might be a promising way to deal with the often reported lack of expertise in the policy process.

To support learning processes, flexibility in financial allocation should be allowed during the project period. In order to broaden the audience benefiting from this social learning, closer involvement of facilitators or moderators working at the interface between researchers and water managers may prove helpful, probably also in the context of commercialising a model. If a model developed is not turned into a commercial product through the involvement of consultancies, post-project funding should be considered to ensure maintenance and ongoing application of the model.

3.3. Managing of expectations-reflections on the role of research

The challenges of the WFD place considerable demands on the water managers for scientific support, especially in addressing the needs of a range of stakeholders. However, the insights that emerged from the Harmoni-CA project strongly suggest that the role of scientists must shift from providers of comprehensive knowledge and information to facilitators of policy processes through the supplemental knowledge that they provide. As Ison *et al.* (2004) state, "*Science and expert knowledge can't reduce complexity of problems. In complex situations, increasing scientific data reveals more uncertainties and new problems and makes visible new interdependencies. Mastering the situation cannot be supported only by scientific data. Social learning helps to make decisions when and because everything isn't known*".

Faced with the limitations in the support that can be provided by specific modelling research projects, water managers can make use of research in a different way. For example, the interface between science and policy should also be strengthened by having researchers act as facilitators or mediators among different stakeholders in participatory processes. Water managers could encourage their stakeholders to engage in participatory modelling processes within research projects using them as a learning platform for their stakeholders. Modellers in turn may find it helpful to seek support in the form of process guidance from experts in public participation to improve the learning processes during model development (e.g., European Commission 2002; HarmoniCOP 2005).

4. CONTENT-RELATED ISSUES FOR A FUTURE RESEARCH AGENDA

Integrated water resources management in a region as diverse as Europe, covers many themes and activities that must be addressed in future research (Gilbert *et al.*, 1998) but will not be discussed within the scope of this Guidance Report. Various documents produced in the Harmoni-CA project confirm the results of the *Common Implementation Process* identifying key issues and research needs under the WFD (de Rooy *et al.*, 2005).

In this report, we have addressed just those themes that were predominant in the workshops and reports. For the reader interested in more details, we recommend the corresponding reports or the Annex of this document which provides relevant extracts of the workshop reports.

4.1. Issues to be addressed by models to support the WFD

Despite the need for a change in the expectations that users have of models as stated above, it was anticipated that additional support for water management to reflect the requirements of WFD would be provided (e.g., in terms of the assessment of measures and integrating indicators). Realizing that this might not be feasible, workshop participants from water management often demanded that research results should provide stronger guidance and a synthesis of the state of the art. However, specific themes were repeatedly identified by water managers. For example, more insights on the interaction of surface and groundwater as well as the stronger linkage of ecological aspects and hydrology in data and models are needed (Borowski and Hare, 2007b; Interwies and Borowski, 2007). In addition, new projects should reflect the integrated objectives of the WFD and other European policies such as the Flora-Fauna-Habitat-Directive or the Flood Directive or the Common Agricultural Policy.

4.2. Economic aspects of the WFD

One of the new challenges of the WFD is clearly the requirement to include environmental resource costs and cost-effectiveness analysis in the decision making procedures for water management measures. The integration of economics has been repeatedly called for. The demand for more research on economic issues is significant and stems from a variety of sources (Hare, 2004, Borowski and Heeb, 2005; Interwies and Borowski, 2007). Methodological issues both in integrated modelling and management, and in economics, are of high importance. Projects were recommended that target specific economic issues such as the comparison of different economic measures in specific basins. At the same time, there was repeated demand for improving the capacity of water authorities by engaging more economists.

4.3. Data availability, accessibility and use in modelling

Harmoni-CA/WP4 has intensively explored the issue of data by looking at the potential for good practice of joint use of modelling and monitoring (Højberg *et al.*, 2010; Refsgaard *et al.*, 2010a).

In the context of this document, it is important to stress that data availability and accessibility including the challenges of appropriate scale was one theme that prevailed throughout all the WP5 workshops. In several of our workshops, and also in the workshops of WP4 focusing on the data issue, a central message sent by water managers to the European funding bodies was the need for alignment of measurements activities and modelling approaches (i.e., no modelling without data collection – no data collection without modelling). The aim is to improve the validation and practical relevance of modelling as well as the focus of data collection (Borowski and Hare, 2007b; Interwies and Borowski, 2007). In the WP5 workshops, there was agreement that many of the most used models in RBM are simple empirical ones for specific management areas and questions. It was suggested that there might however be research and management benefits from trying to extract general rules and lessons from these diverse empirical approaches adopted at the operational management level. It was therefore proposed that an overview of European-wide empirical modelling approaches should be compiled in order to draw out general lessons in using various methods and technical knowledge for river basin management.

5. SUMMARY AND CONCLUSION

Integrated assessment involves both the integration of different scientific disciplines with, and the transfer of research results to management approaches and practices (at the science-policy interface). This Guidance Report has summarized the main insights drawn from the five years of the European Concerted Action Harmoni-CA with emphasis on the challenges of integrated assessment related to the interdisciplinary character of integrated water resources management.

The report clearly shows that structural improvements are necessary to foster the uptake of research in water management. However, it also calls for a distinct perspective on the potential role of models derived through research to support water management. The problems that water managers face can be relatively quickly, but only partially solved by making use of new research results. Water management is a continuous and also cyclical activity that, in the face of greater climatic and other uncertainties, must be open to adaptation.

Another observation of Harmoni-CA is the strong interest by both researchers and water managers to learn from each other. Integrated Assessment is developed and applied over many different research and policy fields (such as climate change, sustainable agriculture, land use, water management, and biodiversity) and within different cultural and institutional contexts methodological approaches used are diverse. As a result of its application to so many fields, research on IA is published in an array of journals and therefore opportunities for scientific exchange are often missed. So the recommendations presented here do not necessarily call for a new science but point to ways in which the exchange and integration of existing knowledge could be improved. For this reason, there is considerable interest in more cooperation among the different scientific communities as well as between water managers and researchers. A new approach to dealing with uncertainty and a new perspective on policy processes is required to improve stakeholder participation in research projects related to water management as well as the practice of water management itself. The scientific community can learn a great deal from the rich experience of the Integrated Assessment community particularly in the context of climate change. Hence more support is needed for the establishment and maintenance of professional networks and exchange forums addressing integrated assessment across the various thematic fields in environmental management. Harmoni-CA can be seen as such an exchange forum and therefore not the last of its sort. Working for European-wide harmonised water resources management is an ongoing endeavour, that will benefit from the experiences and insights of this concerted action.

ACKNOWLEDGEMENTS

This work would not have been possible without all the contributions of the participants in our project activities. We thank them for their valuable contributions and comments during and after the workshops. A special thanks is extended to the Task Force of WP5 for their advice and support: Nils Bull, Carlo Giupponi, Ingo Heinz, Blanka Sárváryné-Szentkatolnay and David Tabara.

6. REFERENCES

Bazzani, G. (in press). Model-based tools regarding the interaction of water management and agriculture. In: Vanrolleghem, P.A. (Ed.) *Decision Support for WFD implementation – Water Framework Directive Series, Volume 3*. IWA Publishing, London, UK.

Borowski, I. and Hare, M. (2007a). Exploring the gap between water managers and researchers: Difficulties of model-based tools to support practical water management. *Journal of Water Resources Management*, **21** (No. 7 – Special Issue "Methods for participatory integrated water resources management"), 1049–1074.

Borowski, I. and Hare, M. (2007b). *Model application by analogy – From technical transferability of model-based tools to transferability by analogy*. Report on 3rd Harmoni-CA/WP5 Policy Workshop. Osnabrueck, Germany, December 4–5 2006.

Borowski, I. and Heeb, J. (2005). *Evaluation of tools/models for integrated & participatory river basin management focusing on the interaction with agricultural management*. Report on 2nd Policy Workshop of Harmoni-CA/WP5, Brussels, Belgium.

Bots, P., Gooch, G., McIntosh, B. and Pahl-Wostl, C. (2010). Understanding the role of perception and valuation in the development and use of models for water management. In: Vanrolleghem, P.A. (Ed.) *Integrated Assessment for WFD implementation: Data, Economic and Human Dimension – Water Framework Directive Series, Volume 2*. IWA Publishing, London, UK. (this volume).

Brouwer, R., Barton, D. and Oosterhuis, F. (2010). Economic methods, models and instruments for the Water Framework Directive. In: Vanrolleghem, P.A. (Ed.) *Integrated Assessment for WFD implementation: Data, Economic and Human Dimension – Water Framework Directive Series, Volume 2*. IWA Publishing, London, UK. (this volume).

Brugnach, M., Tagg, A., Keil, F. and de Lange W. (2007). Uncertainty matters: Computer models at the science–policy interface. *Journal of Water Resources Management* **21**, (No. 7 – Special Issue "Methods for participatory integrated water resources management"), 1075–1090.

de Rooy, M., Broseliske, G. and Menendez, M. (2005). *Common Implementation Strategy for the Water Framework Directive. Key Issues and Research Needs under the Water Framework Directive (Phase 1)*. Policy Summary and Background Document. Environment Directorate, European Commission, Brussels, Belgium.

European Commission (2001). *Common Strategy on the Implementation of the Water Framework Directive*. Environment Directorate, European Commission, Brussels, Belgium.

European Commission (2002). *Common Implementation Strategy for the Water Framework Directive (2000/60/EC)*. Guidance Document No. 8 on Public Participation in Relation to the Water Framework Directive. Environment Directorate. European Commission, Brussels, Belgium.

European Commission (2004). *Common Implementation Strategy for the Water Framework Directive (2000/60/EC). Moving to the next stage in Common Implementation Strategy for the Water Framework Directive. Process and Work Programme for 2005 and 2006*. Environment Directorate, European Commission, Brussels, Belgium.

Gilbert, A., Goosen, H. and Tol, R.S.J. (1998). *Water and Environment. Scoping Study for the European Forum for Integrated Environmental Assessment*. Institute for Environmental Studies, Amsterdam, The Netherlands.

Giupponi, C., Mysiak, J., Depietri, Y. and Tamaro, M. (in press). Decision support systems for water resources management: Current state and guidelines for tool development. In: Vanrolleghem, P.A. (Ed.) *Decision Support for WFD implementation – Water Framework Directive Series, Volume 3*. IWA Publishing, London, UK.

Hare, M. (2003). *Policy Makers' Requirements for Models to Support the Participatory Elements of the Water Framework Directive*. Report on the 1st Policy Workshop of the Harmoni-CA WP5 Project. Harmoni-CA, Osnabrueck, Germany.

Hare, M. (2004). *Modeller's Recommended Research Directions for Models to Support the Participatory Elements of the Water Framework Directive*. Report on the 1st Modellers' & Scientists' Workshop of the Harmoni-CA WP5 Project. Harmoni-CA, Brussels, Belgium.

Hare, M. (2005). *The Use of Models to Support the Participatory Elements of the EU Water Framework Directive: Creating a Dialogue between Policy Makers and Model Makers*. Harmoni-CA. Osnabrück. HCA-WP5-2004-Re04/Final Version.

HarmoniCOP (2005). *Learning Together to Manage Together. Improving Participation in Water Management*. Osnabrueck, Germany.

Hattermann, F.F. and Kundzewicz, Z.W. (2009). *Model-supported Implementation of the Water Framework Directive – A Water Manager's Guide*. IWA Publishing, London, UK.

Heinz, I., Pulido-Velazquez, M., Lund, J.R. and Andreu, J. (2007). Hydro-economic modeling in river basin management: Implications and applications for the European Water Framework Directive. *Journal of Water Resources Management* **21**, (No. 7 – Special Issue "Methods for participatory integrated water resources management"), 1103–1125.

Højberg, A.L., Refsgaard, J.C., Jørgensen, L.F., van Geer, F. and Zsuffa, I. (2010). Good practise in joint use of monitoring and modelling. In: Vanrolleghem, P.A. (Ed.) *Integrated Assessment for WFD implementation: Data, Economic and Human Dimension – Water Framework Directive Series, Volume 2*. IWA Publishing, London, UK, (this volume).

Horlitz, T. (2007). The role of model interfaces for participation in water management. *Journal of Water Resources Management* **21**, (No. 7 – Special Issue "Methods for participatory integrated water resources management"), 1091–1102.

Interwies, E. and Borowski, I. (2007). *Using Economic Methods and Models for the Implementation of the Water Framework Directive: Status, Options and Challenges for a Better Integration of Water Economic Research into Policy Implementation*. Report on the 4th Harmoni-CA/WP5 Policy Workshop, Berlin, Germany, June 14–15.

Ison, R.L., Steyaert, P., Roggero, P.P., Hubert, B. and Jiggins, J. (2004). *Social Learning for the Integrated Management and Sustainable Use of Water at Catchment Scale*. SLIM final report, Open University, Milton Keynes.

Newell, B., Crumley, C.L., Hassan, N., Lambin, E.F., Pahl-Wostl, C., Underdal, A. and Wasson, R. (2005). A conceptual template for integrative human–environment research. *Global Environmental Change* **15**, 299–307.

Pahl-Wostl, C. (2002). Towards sustainability in the water sector – The importance of human actors and processes of social learning. *Aquatic Sciences* **64**, 394–411.

Pahl-Wostl, C. (2007a). The implications of complexity for integrated resources management. *Environmental Modelling & Software* **22**, 561–569.

Pahl-Wostl, C. (2007b). Transitions towards adaptive management of water facing climate and global change. *Water Resources Management* **21**, 49–62.

Refsgaard, J.C., Jørgensen, L.F. and Højberg, A.L. (2010a). Data availability and accessibility: in view of model-based water management. In: Vanrolleghem, P.A. (Ed.) *Integrated Assessment for WFD implementation: Data, Economic and Human Dimension – Water Framework Directive Series, Volume 2*. IWA Publishing, London, UK, (this volume).

Refsgaard, J.C., van der Sluijs, J.P., Højberg, A.L. and Vanrolleghem P.A. (2010b). Uncertainty analysis in model-based water management. In: Vanrolleghem, P.A. (Ed.) *Modelling Aspects of Water Framework Directive Implementation – Water Framework Directive Series, Volume 1*. IWA Publishing, London, UK.

Rotter, S. and Rasche, K. (2007). *A telephone survey – based assessment of twelve research projects*. Report on end user involvement in research projects. Harmoni-CA. Available at www.harmoni-ca.info/products.

van Bers, C., Petry, D. and Pahl-Wostl, C. (Eds.) (2007). *Global Assessments: Bridging Scales and Linking to Policy*. Report on the joint TIAS-GWSP workshop held at the University of Maryland University College, Adelphi, USA, May 10–11 2007. GWSP Issues in Global Water System Research, No. 2. GWSP IPO, Bonn. Available at: www.gwsp.org

Vanrolleghem, P.A. (Ed.) *Modelling Aspects of Water Framework Directive Implementation – Water Framework Directive Series, Volume 1*. IWA Publishing, London, UK.

7. ANNEX: RESEARCH DIRECTIONS IDENTIFIED IN HARMONI-CA/WP5 WORKSHOPS

Participatory research processes and research directions from 1st Methods Workshop

(Extract from Hare M. (2004): Modeller's Recommended Research Directions for Models to Support the Participatory Elements of the Water Framework Directive. Report on the 1st Modellers' & Scientists' Workshop of the Harmoni-CA WP5 Project. Harmoni-CA. Available at www.harmoni-ca.info)

Over 30 recommended directions suggested by the participants have been categorised according to whether they lie in the areas of human dimensions research, participatory research, or modelling design research. The tables on the following pages summarize the research directions that were identified at the workshop. It will be important for readers to identify which areas already have programmes running and which do not.

The directions have been categorised according to 3 categories and associated sub-categories:

- Human dimensions (HD) – how different human dimensions can be calculated, how data can be elicited and how they can be modelled (Table 1)
- Participation support – how participatory modelling processes can be optimised, supported, or evaluated (Table 2), with the sub-catogories
  - Evaluation of participation (PE)
  - Stakeholder analysis (PSA)
  - Workshops and facilitation (PW)
  - Participatory modelling (PM)
- Modelling design – issues directly to do with design of models (Table 3), with the sub-categories
  - DPSIR (MD) – how this specific methodology can be improved
  - Interfaces (MI) – how interfaces can be designed for participatory use
  - Scale (MS) – how scales issues can be modelled and dealt with
  - Tailoring (MT) – how and whether models can be tailored
  - Model use (MU) – how models can be used

The participants were asked during the final session to mark three key research directions, from day two, that they thought were important. The tables below therefore also show the total number of votes for each day two research direction.

Table 1 Key research directions in Category 1: Human dimensions research actions (no sub-categories)

Key Research Directions	votes
defining linkages between socio-economics and hydrology	7
integrate human dimensions	3
database of existing socio-economic projects/ models	1
coupling models	1
specifying and defining human dimension	0

Table 2 Research direction of Category 2: Participation support including sub-categories

Sub Category	Research direction	votes
Evaluation	matrix for efficiency and ongoing evaluation of process	3
	cost effectiveness of pp/pre-assessment – case studies based	2
Stakeholder analysis	pitfalls of selecting representatives from different groups – how – what it means	2
	role playing as an analytical tool for stakeholder identification and selection plus character of stakeholder/ how well will they contribute?	2
	linking spatial info at different scales to identify stakeholders	1
	institutional capacity and trust analysis	
Workshops and facilitation	how to use (scenario) workshops for participatory river basin modelling	2
	methodologies for maintaining stakeholder participation	1
	how to reduce high expectations during the modelling process/ how to show potential for decision making	
Participatory Modelling	participatory modelling – what are the best forms, how can it be done?	n/a
	case study collections for participatory modelling for policy makers	n/a
	how and when is public and stakeholder use of tools to be included in planning?	n/a

Table 3 Key Research Directions from Category 3: Model design, including sub-categories

Sub Category	Research area	votes
DPSIR	best practice/ cases/ narratives for participatory modelling in scenarios/DPSIR etc	2
	communication of causal links in DPSIR	
	improving dynamic feedbacks in DPSIR	
Interfaces	ergonomics of interfaces for participatory river basin modelling	1
	presentation displaying information about risks – the art of	
Scale	upscaling of data for IAM	1
	upscaling of scenarios & multi-scale scenarios	
	transboundary catchment models	n/a
	multi-model approach for dealing with different temporal and spatial scales	n/a
Tailoring	tailor-made models for different stakeholders/ is it necessary to adapt the tools for communication?	2
	guide or expert system allowing managers to tailor a participatory multi-tool approach for their own purposes	2
	general vs tailor-made modelling systems/ flexibility/misuse – which is better when?	
Use	field experience in cyclic modelling	2
	how can models support a fair process in conflict situations?	
	toolbox of techniques and applicability of models supporting exchange of views	
	how do you combine the use of different tools and models in participatory management?	n/a

Recommendations of 2nd Harmoni-CA/WP5 policy workshop on models supporting the interaction of agricultural and water management

(Extract from Borowski I. and Heeb J. (2005). Evaluation of tools/models for integrated & participatory river basin management focusing on the interaction with agricultural management. 2nd Policy Workshop of Harmoni-CA/WP. Available at www.harmoni-ca.info)

Supporting participation
The existing tools/models can support and facilitate participation, but stakeholders have to be involved more intensively in the tool/model development process to bridge the knowledge gap between them, the "river basin managers" and the scientists (or in other words to make better use of existing knowledge).

Supporting economic assessments
There is a clear need for tools/models to support integrated economic assessments of policy measures. Existing tools/models should be integrated with existing agro-economic tools/models.

Transferability and applicability
Models should be developed to make a region-specific evaluation of policy measures possible. This however means that rather than having a few general tools/models there is a need for a large number of specific, regional tools/models; which in turn requires more resources. Tools/models should facilitate conflict and synergy management between the Common Agricultural Policy (CAP) and the WFD.

Integration
Tools/models should be developed to make more systemic work possible, integrating ecological, economic and social issues on a qualitative and quantitative level.

Simulation of behaviour
Tools/models are needed to simulate farmer's behaviour in response to policy measures. This would help to work out concrete action plans and select, based on the simulation results, the most sustainable scenario.

Motivation policy makers to use the tools/models
Tools/models are still mostly used in the academic world. In order to convince policy makers to use tools/models in their daily work information about the abilities, the validity, the level of uncertainty and the conditions of applicability of the tools/models have to be clearer to the practitioners. Tools/models should make clear what added value they can provide to policy makers working

in integrated and participatory river basin management in the CAP – WFD context. The access to the tools/models should be made easier for the policy makers: A promising option are toolboxes as developed. in the BMW project (see http://www.ymparisto.fi/default.asp?node = 11687&lan = en for more information on project) and the Harmoni-CA project (www.harmoni-ca.info). In addition to technical solutions more work needs to be done on understanding the constraints of policy makers so that tools/models can be adapted to fit their needs.

Table 4 presents the list of activities within participatory river basin management to be initiated or taken further as identified by workshop participants. It can be seen as a "To-Do-List". The red arrows represent the required "distance to go" (as interpreted by the authors of this report) to develop the existing tools/model from the actual state (left hand side of the arrow) to the target state (right hand side of the arrow).

Table 4 Visualization of "distance to go" to meet the key requirements of the participants

Key requirements of the participants	-- Requirement not met	- Requirement partially met	+ Requirement nearly met	++ Requirement met
Supporting participation			——	→
Supporting economic assessments		——	——	→
Transferability and applicability		——	——	→
Integration		——	——	→
Simulation of behaviour	—	——	——	→
Motivating policy makers to use tools/models		——	——	→

Note: The use of models to support participatory river basin management is rare. This is due not only to the fact that models are not used by policy makers to a great extent but also because participation, as it is viewed by the scientific community, is not yet a common activity in policy making.

Summary and Conclusions from 4th Harmoni-CA/WP5 Policy Workshop: Elements of a future research agenda

(Extract from Interwies E. and Borowski I. (2007). Using Economic Methods and Models for the Implementation of the Water Framework Directive: Status, Options and Challenges for a better integration of research on economic aspects of water into policy implementation. Workshop report on the 4th Harmoni-CA/WP5 Policy Workshop. Berlin, Germany, June 14–15 2007. Harmoni-CA, available at www.harmoni-ca.info)

Based on the presentations at the workshop, the discussions in the plenary, and the results of the working group sessions, a number of topics for further research were identified. These include content-related research and the improvement of the research-policy interface. The results also revealed that when focusing on economic issues, the challenges for the science-policy interface are often of a rather general nature. They start already during the definition of the problem- for which water managers need solution and on which scientists what to do research on. The subtleties of the related challenges were discussed more deeply in Borowski and Hare (2007a).

Content related issues of a future research agenda:
Demand for more research on economic issues was broad and diverse. Asked for prioritization (see section 2), the participants maintained this diversity by spreading their votes relatively even between the different issues. However, more research on integration and integrated models was valued highest, directly followed by more research on economic methods and tools in general and more clarity on definitions of main terms regarding WFD-economics.

Within the continuum of economic science "interference" to the political decision making, three levels of this interference were identified: At a first level, economic science can inform the political decision-making process, while also the limitations of the approaches used need to be made transparent in order to reduce the risk of "backfiring"/misuse of economic assessment. Using economics for examining/assessing problem-solving approaches (regarding effectiveness but also distribution) needs to be dealt with cautiously, since the risk of misuse here is given. At a second level, beyond informing decisions transparently by scientific approaches applied in the WFD implementation, transparency was also repeatedly discussed in the context of public participation and negotiation with/ between different stakeholders. Scientists were also expected to act as a "transmitter" of knowledge beyond policy and as a facilitator to improve processes of joint implementation and social learning. At a third level, economics should not be used for legitimizing already taken decisions; which would indicate a process failure in the overall WFD implementation.

The results on the content issues might indicate two things: Methodological issues both on integrated modelling and management, and on economics, are of high importance. The diversity of the content of the research demanded (e.g., on scales, uncertainty or case studies) indicated that especially content depends often more strongly on the specific situation, not only in terms of basin specifications and national/ regional situation but also in terms of available capacity. Although not highly ranked in the prioritization exercise, data availability & accessibility including the challenges of appropriate scale, was one theme which was persistent throughout the workshop. Harmoni-CA has explored this issue of data intensively by looking at the good potential practice for joint use of modelling and monitoring (Højberg *et al.*, 2010; Refsgaard *et al.*, 2010b).

In the following, the main research topics identified from the discussion are listed, organized along the working group themes.

Research demand on the selection of measures, cost-effectiveness-analysis, integrated (hydro-economic) modelling:

- Approaches supporting systematic assessments of uncertainty throughout the dose-response chain are needed, including environmental, social and economic system responses to measures;
- Understanding of the value of extra information to reduce uncertainty in decision-making (through Bayesian approaches should be improved;
- Integrated-extended CEA approaches (versus CBA and MCA), including weighting procedures for multiple objective 'satisfying' problems/policy contexts; should be tested.
- Integrated modelling, better linking biophysical to economic models for different purposes at different scales (e.g., adaptation-mitigation); should be improved.
- Development of standards for non-market benefit analysis (based on the Aquamoney-project) was required.

Research demand on cost-benefit analysis, exemptions, and environmental benefits:

- Better understanding of "disproportionality of costs" (at strategic/ national level; local implementation) and to which extent a CBA is needed;
- Research on the role of non-monetary benefits: what & how to assess?
- Increase the applicability of valuation approaches: more specific non-market valuation studies (of specific WFD-relevance); methodological work on benefit transfer;
- Work on how to better conduct market-based benefit evaluation (e.g., at a bigger scale);
- Better understanding and assessment of opportunity costs;
- Analysis of stakeholder acceptance and exploring ways to increase it;
- Role of monetary valuation/CBA in general in decision making.

Research demand on water pricing, cost recovery/polluter-pays principle, economic instruments:

- More case studies are needed: to learn from approaches in other countries but also to test new instruments and approaches coming from research;
- More studies needed on a broader definition of water services including e.g., hydropower, navigation etc.;
- More research is needed to find the appropriate criteria for cost recovery, polluters-pays-principle and cost distribution, including aspects such as affordability, hidden costs (e.g., subsidies), spatially differentiation of approaches, and justification of exemptions;
- How to take the results from small scale studies and use them on a larger scale?
- Environmental Resource Costs: research on practical implementation, and on quantification needed providing guidance to water managers;
- Continue research on tradable markets, e.g., nutrient trading.

Guidance Report II.2

Data availability and accessibility in view of model-based water management

Jens Christian Refsgaard, Lisbeth Flindt Jørgensen and Anker Lajer Højberg

Preface

This Guidance Report is produced within the framework of Harmoni-CA – Harmonised Modelling Tools for integrated River Basin Management. Harmoni-CA was a concerted action supported by the European Commission under the Fifth Framework Programme and contributing to the implementation of the Key Action "Sustainable Management and Quality of Water" (EKVI-2001-00192). Harmoni-CA aimed at facilitating the dialogue and helping to bridge the gap between research and policy, by synthesizing available knowledge produced by the various RTD (CatchMod) projects and facilitating the development and the use of these methodologies and tools to support the use of Information, Communication and Technology (ICT) tools in implementing the Water Framework Directive (WFD).

Harmoni-CA work package 4 (WP4) worked with "Joint use of Monitoring and Modelling", especially in relation to implementing the WFD. In this context it has been found relevant to give an overview on the status of data in terms of availability and accessibility; both in relation to the implementation of the Directive and in relation to related research activities.

To elaborate the issue of joint monitoring and modelling five workshops were organized between 2004 and 2007. Common frames for the workshops have been that it was essential to attract the right people; both with respect to researchers capable of bringing forward ideas in an easy understandable way and when it comes to practitioners representing the water management world. This has not been an easy task, especially as most national practitioners are not used to attending international conferences or workshops conducted in English. Thus the first bottleneck was to identify the relevant persons and persuade them to attend the workshops. We would like to thank the about 80 people for their contributions to our workshops. The 80 participants came from 21 of 27 EC countries plus Norway, Moldava, Ukraine and the US. The balance between researchers/modellers and water managers were equal, a few representing both sides. Besides, a couple of consultants attended the workshops, mostly presenting tangible examples from all around the world. Several attended more than one workshop. Experiences, conclusions, recommendations and lessons learned are brought forward in this report.

The information in the sections on data availability and previous modelling studies in the eight selected river basins have been reviewed and modified by the following HarmoniRiB partners: Technical University of Crete, Chania (Ioannis Tsanis); Universidad de Castilla la Mancha, Albacete (Salomon Montesinos); Water Research Institute CNR-IRSA, Bari (Guiseppe Passarella); Institute of Inland Water Management and Waste Water Treatment, RIZA, Lelystad (Michiel Blind); Centre for Environmental Research, UFZ, Magdeburg (Michael Rode); Centre for Ecology and Hydrology, Wallingford (Roger Moore); Povodi Moravy, Brno (Pavel Biza); and Geological Survey of Denmark and Greenland, GEUS, Copenhagen (Jens Christian Refsgaard). Their support is gratefully acknowledged.

Earlier versions of this document have been reviewed by Dr. Bernardas Paukstys, Water House, Vilnius, Lithuania, Mr. Morten Sørensen, Environment Centre Odense, Denmark and Dr. Faycal Bouraoui, Joint Research Centre, Ispra, Italy and discussed at the Harmoni-CA Review Workshop held in Köln 28/02 – 02/03/2007. The final document has benefited from comments from the participants at the fifth Harmoni-CA WP4 workshop held in Budapest 20/06 – 21/06/2007 and from Dr. Wim de Lange, RIZA, The Netherlands and Dr. Ilke Borowski, University of Osnabrück, Germany.

Keywords

Databases, Mathematical Modelling, Monitoring, River Basin Management, Uncertainty

EXECUTIVE SUMMARY

Data for modelling – what are the problems?

The need to carry out monitoring = collecting data – is emphasised both in the legal text of the WFD and in the supporting Guidance Documents. Existing data are needed to characterise and assess the status of water bodies and new data are needed to verify or to follow the development in that water body. Thus, a lot of data collection is carried out to fulfil the requirements of the WFD – but how accessible are these data in practise and are there enough data to support modelling?

The question on how much data are required for modelling is classical. In the modelling community lack of data is often used as an 'excuse' for poor model performance. Similarly, lack of data is often used by water managers as a reason for not applying models. On the other hand, it is always possible to set up a model with whatever sparse available data, but the uncertainty of the model result might easily be far too high. Therefore, the question should be rephrased as: "*Is the model prediction uncertainty small enough to be of any practical use in a water management context*?" This question cannot be answered without considering the socio-economic context – that is how much is at stake economically and politically for stakeholders and decision makers. And this answer will vary from case to case.

Data availability

There is a large diversity of data availability in river basins across Europe. For some data types like meteorological and river discharge data the coverage only varies moderately among river basins. When it comes to data on surface water quality and biology and on groundwater quality there are, however, very large differences. This difference appears to a large extent to reflect the seriousness of the pressures (population density, pollution, etc.); however, the water management traditions and the socio-economic policy priorities may also be important factors.

The level of complexity of modelling carried out in river basins appears in general to be proportional to the amount of data available. This may be because both the amount of data and the need for modelling to a large extent is driven by the level of pressure on the water resources, i.e., on the perceived water management problems. So maybe, at least in areas and domains with severe problems, the interaction between data and modelling is in fact larger than we think even if it many times may not be explicitly recognised by all involved actors.

The analysis therefore suggests that the larger problems at hand, the more data are available, and the more modelling studies have typically been carried out. It can therefore be concluded that data availability is seldom the largest constraint for not doing modelling.

Analysis of WFD Article 5 reports suggests that data availability in many cases may not be sufficient for all aspects of the coming WFD implementation. This implies that the level of modelling required to support the WFD implementation will be somewhat higher than what can be performed with the present data availability. So, although the present data availability appears to be adequate for some level of modelling, it may in many cases not be sufficient for the level of modelling required from a socio-economic (water management/ WFD) point of view.

Data accessibility

One question is whether data are actually being collected – the next crucial question is whether the collected data are stored efficiently and are easily accessible to all potential users.

Several European and International conventions such as the Aarhus Convention and the INSPIRE Directive prescribe free data accessibility. Besides the EC has developed a web portal "WISE" for data collected in connection with the WFD. However, in addition to the data collected specifically for the WFD much more data are being collected in the river basins. These data that are useful for water management purpose can be difficult to access. The most important constraints in this respect are:

- Economical: Some institutes have to charge a fee to cover costs related to data collection.
- Political: Conflicting interests among data collectors and/or data users can hamper the access.
- Data formats: Most data exist in digital formats but even so are not harmonised. Furthermore, some data are still only available on paper.
- Fractionised databases: Often data exist in many databases with different data owners with different data formats and dissemination policies.
- Transboundary barriers: There still seem to be problems in harmonising or exchanging data in river basins that cross national borders.

These bottlenecks represent considerable constraints in daily water management. But also in research the problems exist in getting data that are collected but not easily accessible – often it requires a lot of effort both in terms of costs and hard work.

In conclusion there still seems to be a large need for pan-European policies on making data collected using public funds free for all.

Data quality – uncertainty

In addition to the availability and the accessibility of the data, the quality of the data is crucial. If data are not trustworthy they are not useful. Understanding the uncertainty in environmental data and systems is essential for making robust and wise water management decisions. Therefore, there is a need to characterise the data quality in terms of uncertainty and store such information as part of the data documentation.

Data for research purposes

Data requirements for research purposes vary tremendously depending on the objective of the research study. As the nature of research is to gain new knowledge, there will seldom be sufficient existing data on the particular research topic. If, for instance, the objective is to improve the understanding of water flow and reactive transport processes at a certain scale in a given area then it will most often be required to design a dedicated field monitoring programme just for the research project. If, on the other hand the research is focussing on water management issues and decision making there will often be sufficient data on the hydrological system, but instead it may be required to collect information on e.g., stakeholder views and behaviour.

In research many resources go into acquiring data for specific research projects – data that may only be available for individual projects. Thus there is a need to establish networks of representative river basins and hydrological observatories with data freely available for research purposes.

1. DATA FOR MODELLING – WHAT ARE THE PROBLEMS?

1.1. Why is modelling an important tool in water resources management

The basis for good water resources management is adequate data implying that monitoring is a prerequisite for any work related to the implementation of the Water Framework Directive and the Groundwater Directive. Data alone do, however, only provide information of the current state of the environment, e.g., a chemical concentration, at a specific location and at the time at which the data was collected. This information is, in itself, of very little use. If more measures of the compound have been taken a time series can be constructed and obvious trends in the concentration levels may be revealed. The data do, however, not explain the cause for this trend. To extract this kind of information from the data set one needs to interpret the data based on an understanding of the physical system, that is, we formulate a conceptual understanding (or conceptual model) of the system from which we explain the data. As such we add knowledge to the data, which is necessary if we want to formulate future management strategies based on observations. Natural systems are, however, very complex with several processes occurring simultaneously and interacting. It is therefore very difficult to fully comprehend and separate the different processes, their interaction and their consequences by interpretation of field data alone. In this context, models can be seen as a formal description of knowledge of the hydrological system studied, describing the spatial and temporal coherence in the system.

Monitoring data and models provide complementary information of the system. The model captures physical knowledge about the systems behaviour and monitoring provides information from the actual system including phenomena which are not included in the model. It may be argued that "modelling without monitoring data is guesswork" and "monitoring without models is a waste of money".

Modelling and monitoring activities can support each other in many ways. On the one hand useful modelling is not possible without appropriate data that must originate from monitoring. On the other hand modelling can support monitoring activities through different activities such as (a) quality assurance of monitoring data; (b) knowledge-based interpolation and extrapolation in time and space; (c) establishment of conceptual models with causal relationships; (d) assessing effects of anthropogenic activities e.g., in relation to programme of measures; and (e) designing of monitoring systems (Højberg *et al.*, 2007).

1.2. Objectives and target audience of this document

The overall objective of the present report is to analyse and discuss state-of-the-art on existing data in relation to model applications for research purposes and for implementation of the Water Framework Directive and the Groundwater Directive. More specifically the objectives are:

- To assess the presently available data in a number of European river basins.
- To assess the accessibility of existing data.
- To provide recommendations on data availability and accessibility for WFD applications.
- To provide recommendations on data availability and accessibility for WFD relevant research studies.

The report is not intended to be a research report, but rather a synthesis of existing knowledge and experience as it has emerged during the series of workshops held within the Harmoni-CA project.

The target audience is policy makers, water managers and professionals involved in implementation of the Water Framework Directive and the Groundwater Directive.

1.3. Water Framework Directive – requirements

In the process of implementing the Water Framework Directive (WFD) the issue of data is extremely important. The WFD asks for ambitious goals for the (water) environment with the ultimate objective to obtain good quantitative and qualitative status of all waters in Europe by 2015. Achieving this calls for the solution of a large number of technical tasks, all involving either using already collected data or collecting new data sets: water bodies need to be delineated, classified and possibly grouped; water districts must be characterised, anthropogenic effects evaluated, trends identified, pressures and impacts assessed, programmes of measures established and evaluated, etc.

The WFD is supported by a number of Common Implementation Strategy (CIS) Guidance Documents (GD) of which the most relevant in connection to data is the Monitoring GD (EC, 2003). This GD prescribes 4 types of monitoring:

- *Surveillance Monitoring*: To determine status; to supplement and validate the initial impact assessments; to assess long time changes; to validate hypotheses and to assist in design of future monitoring programmes.
- *Operational Monitoring*: To assess status of water bodies at risk of failing to meet environmental objectives; to assess effects of programmes of measures. Operational monitoring must be carried out in all bodies receiving priority substances. The water bodies may be grouped.
- *Investigative Monitoring (only surface water)*: Where the reason for any exceedance is unknown; where good status is not likely to be achieved and where operational monitoring has not already been established; to ascertain the magnitude and impacts of accidental pollution; Measure only the problem-defined variable(s).
- *Water Level Monitoring Network Design (only groundwater)*: For groundwater quantitative monitoring is required. This should be designed to ensure that the available groundwater resource is not exceeded by the long-term annual average rate of abstraction; to ensure that abstractions and other anthropogenic alterations to groundwater levels do not adversely affect associated surface water bodies and terrestrial ecosystems that depend directly on groundwater with respect to their water needs; to ensure that anthropogenic alterations to flow direction must not have caused, or be likely to cause, saltwater or other intrusion.

As the WFD and the supportive CIS Guidance documents have a generic approach, they do not necessarily provide extensive lists of data or explicit recommendations on spatial and temporal resolution etc. (Blind and de Blois, 2003). Further they do not give explicit recommendations on how to store data in order to fulfil international conventions prescribing freely available data (UNECE, 1998; EC, 2007). A more detailed discussion of the data requirements for the WFD implementation is given in Section 2.2.

In addition to the data from the WFD monitoring the WFD planning process that will be completed with the preparation of River Basin Management plans require a variety of additional data, most of which typically already exist. These additional data include system data describing physical and chemical characteristics as topography, river geometry, soil types, soil hydraulic properties and geology. Finally, more detailed data than formally requested in the WFD monitoring are often available from stakeholders such as water supply companies, agriculture, industry and citizens. All these additional data are crucial for modelling studies.

The new water management requirements posed by the WFD call for the development and application of new tools and methods. Comprehensive European and national research programmes have been initiated to support this need for new operational approaches. Many of the research projects funded in this connection require a substantial amount of data, both existing data and new types of data. Research projects collecting existing data are often faced with imposed conditions restricting their ability to make the data freely available to other users after project completion.

In this context it is clear that already collected data will play an important role at the same time as the need for collecting new data is obvious. Thus state-of-the-art on data collection is very relevant in terms of the following issues:

- What sort of data is already being collected?
- What type of modelling can be carried out with these data? Is this sufficient for WFD implementation?
- Are the existing data sufficient for research studies?
- Are data easily accessible?
- Are costs constraining?

1.4. Do we have enough data for modelling?

The question on how much data are required for modelling is classical and has been debated at length during several decades. In the modelling community lack of data is often used as an 'excuse' for poor model performance. Some modellers claim that lack of data or lack of quality in data is the main constraint for practical use of existing modelling tools (Steenstra *et al.*, 2004; Kamphorst *et al.*, 2005) and for progress in development of improved process descriptions for models (Beven, 1996). Additional data will often improve the performance of models, and therefore it can be argued from a scientific point of view that there is always scope for improvements and there will never be data enough.

Similarly, lack of data is often by water managers used as a reason for not applying complex models, but instead using simple tools (Steenstra *et al.*, 2004).

A general discussion on the adequacy of data for modelling is given in Section 2.3 and other sections in Chapter 2. More specifically, the issue of data availability for modelling is discussed in Chapter 3.

1.5. Are the existing data accessible for modelling?

Using and extracting information from already collected data is often the first step in water management. This requires that the water manager can acquire the existing data within the area of interest. But the task is not finished when the data are identified. Existing data are not always easily accessible to interested users. The five main constraints are:

- Economical – cost of data.
- Political barriers.
- Many different data formats.
- Practical problems in transboundary co-operation.
- Fractionised databases.

Further discussion of data accessibility is given in Chapter 4.

2. DATA REQUIREMENTS – METHODOLOGICAL FRAMEWORK

2.1. Modelling categories

There is a huge diversity of model types and purposes of modelling, as well as of hydrological regimes. Therefore, it is not straightforward to specify how much data are required for modelling. When we in the following try to do so anyway we do it with the objective of having a framework for analysing and discussing data availability in relation to modelling and we should like to emphasise that it must by no means be seen as standards prescribing how much data are required. Thus, the attempt made in the following does not pretend to provide universally applicable results and many examples for which it does not apply can no doubt be found.

We characterise modelling in two dimensions: (a) the level of complexity; and (b) the modelling domain.

The level of complexity is subdivided into four categories inspired by the classification of hydrological models into empirical, lumped conceptual and distributed physically-based (Refsgaard, 1996):

- *Simple models – Screening*. By this we understand model codes with relatively simple process descriptions. This does not imply that they automatically are simple to understand and operate. Simple models are typically based on regression or empirical equations and are therefore often referred to as *empirical or black box models*. In these models the cause-effect relation is typically established based on analysis of concurrent input and output time-series analysis, and not from an understanding of the processes of the physical system. Provided an input (driving force) the models can be used to estimate the output response, but only for the variable for which a relation has been established. Empirical models typically require long time series to establish the input-output relations specific to the studied site, but once the relation has been defined the data requirement is generally limited to the input data. The requirement to system data is usually very limited. Simple models are today often applied in a user-friendly GIS environment. They typically require relatively less amount of data in operational use and are therefore quick to use. They are useful for screening purposes, where the requirements to prediction uncertainty are limited.

- *Intermediate models – Planning*. Models in this category are based on a conceptual understanding of the system and processes (*lumped conceptual models*). The descriptions are, however, lumped where all parameters and variables represent average values. Hence, the equations of the lumped models are semi-empirical, but still with a physical basis. As the models are based on a conceptualisation of the system long time series are not needed to establish an interrelation between the variables. Data are, however, required for model calibration and system data are needed to derive some aggregated characteristics of the model domain. These models are well suited for planning purposes in cases where accuracy needs to be better than for initial screening of problem areas and where some moderate level of extrapolation beyond existing situation and data are required.
- *Comprehensive models – Design*. These models rely on physical laws, and the flows of water and energy are calculated directly from the governing continuum (partial differential) equations. By a distributed description in time and space, these models provide a more detailed description of the hydrological system (*distributed physically-based models*). Furthermore, models in this category are able to make use of all available data from the system, and provide estimates of all internal state variables, not only those of primary focus. Compared to the other model types, the distributed models generally require a more detailed spatial resolution of data, which increases the requirements to system data. These models are often applied to help design management actions, e.g., by scenario analysis of alternative options for changes in land use, agricultural practice, water abstraction or remediation schemes.

- *Process studies – Research.* Research studies aiming at obtaining better insight into physical processes or testing data measuring instruments or new data types are usually supported by modelling analyses. These models can be based on existing model codes, often of the comprehensive, physically-based type, but can also be new codes developed particularly for the specific research study.

It must be emphasised that this classification is schematic and that many model codes or applications do not fit exactly. Some model codes fall naturally into one of the above four boxes, while others may be placed in more than one. Similarly, model codes that are classified as simple are some times applied also to planning and some times even design purposes.

The modelling domain can be subdivided into the following six groups:

- *Water resources assessment* is the determination of the availability of surface water and groundwater in space and time, including the impacts of the various water uses. Model types include catchment scale models for water allocation, rainfall-runoff, groundwater flow and integrated/coupled versions of these.
- *Floods* are here concerned with modelling for assessing extents and risks of flooding and for assessing flood control schemes. These models are typically 1D/2D hydraulic/hydrodynamic models of rivers and flood plains, maybe with rainfall-runoff models for providing inflow hydrographs.
- *Surface water quality and ecosystems* is concerned with studies of water quality and ecological aspects in rivers, lakes and estuaries.
- *Agricultural management*, including non-point pollution often deals with field scale studies of irrigation, fertiliser application and leaching and pesticide application and leaching. The most comprehensive of these models also includes simulation of crop yield as a consequence of the applied agricultural management practise.
- *Groundwater quality* is the study of groundwater vulnerability to pollution and the transport and fate of contaminants through the aquifer systems.
- *Socio-economic cost assessment* is the social and economic assessments of various actions considered to improve the ecological condition towards achieving good water status.

Typical types and examples of model codes for the different categories are listed in Table 2.1.

Table 2.1 Types of model codes and examples of codes suitable for different domains and different levels of model complexity

Complexity of code – Application *(Data Requirements)*	Water Resources Assessment	Floods	Surface water quality and ecosystems	Agricultural management, including non-point pollution	Groundwater quality	Socio-economic cost assessment
Simple models – Screening *(Low)*	GIS based water balance tools, MIKE BASIN		Regression models, Vollenweider-Model, Moneris	Regression models, Export coefficient models, USLE	DRASTIC	Cost effectiveness
Intermediate models – Planning *(Intermediate)*	Sacramento, HBV	HEC-RAS 1D, Lisflood		HBV-N, INCA, AGNPS,		Hedonic pricing method, Contingent valuation method, Cost-benefit
Comprehensive models – Design *(High)*	MIKE SHE, MODFLOW	MIKE 11, SOBEK, ISIS, Telemac	DELFT3D-ECO WASP7, RWQM1, Aquasim	SWAT, DAISY, SOIL-N, ANIMO WEPP, EUROSEM	PHREAQC, HST3D, RT3D	
Process studies – Research *(Very high)*	Comprehensive model codes + tailor made research codes	Comprehensive model codes + tailor made research codes	Comprehensive model codes + tailor made research codes	Comprehensive model codes + tailor made research codes	Comprehensive model codes + tailor made research codes	Comprehensive model codes + tailor made research codes

2.2. Data requirements for the WFD implementation

The data requirements in a river basin depend on many factors. Fundamentally it depends on the economic and social/political value of the data, which again depends on the water use and conflicting demands in a socio-economic context and on the hydrological regime. As noted by WMO (1994) it is in practise impossible to design a monitoring network from purely objectively scientifically based criteria taking all relevant aspects into account. It is therefore also impossible to prescribe universally applicable data requirements for water management, including modelling.

One point of departure in search for guidance on data requirements could be WMO's Guide to Hydrological Practices WMO (1994) which outlines requirements to density of stations for minimum network depending on physiographical regions. For interior plains and hilly/undulating regions the minimum densities of gauging stations are specified as:

- Precipitation (non-recording, daily): 575 km^2 per station.
- Evapotranspiration: 50,000 km^2 per station.
- Discharge: 1875 km^2 per station.
- Sediment transport: 12,500 km^2 per station.
- Surface water quality: 37,500/45,000 km^2 per station for the two regions respectively.

These requirements are seen to be very/extreme minimal and intuitively far below what is required in a WFD context. These minimum requirements would result in only one surface water quality gauging station for countries such as The Netherland and Denmark. With the recognised eutrophication problems in these countries this is obviously far from sufficient and as a fact much more stations actually exist.

Another place to look for recommendations on data requirements would be the CIS Guidance Documents. This has been done by Blind and de Blois (2003). Data requirements for the WFD are discussed in many of the WFD guidance documents, e.g., those dealing with 'Monitoring' (EC, 2003b) and 'Analysis of pressures and impacts' (EC, 2003a) as well as 'Planning Processes' (EC, 2003c). In the monitoring guidance document (EC, 2003b) some general guidance is provided on where, what and when to monitor in the different monitoring programmes. Most detailed information is provided on which variables that must be included in the monitoring programmes, while the required temporal and spatial resolution in general are linked to the chosen level of confidence and precision, where a key principle is stated: "*the actual precision and confidence levels achieved should enable meaningful assessments of status in time and space to be made. Member States will have to quote these levels in River Basin Management Plans and will thus be open to scrutiny and comment by others*". The acceptable level of precision and confidence is thus a subjective quantity that depends on the socio-economic interests that are at stake and of the risk strategy of the decision makers (e.g., Bots *et al.*, 2010). Therefore, there is a balance between the costs of monitoring against the potential risk and, in general, the lower the desired risk, the more monitoring is required.

The monitoring guidance (EC, 2003b) does, however, recognise that the required resolution depends on both the dynamic of the variables and the complexity of the system. For the temporal resolution a frequency must be chosen that allows the detection of both short- and long-term changes of the physical system. Highly variable parameters should therefore be measured more frequently than parameters that are more stable or relatively unresponsive to short-term variation in stresses and pressures. Furthermore, the monitoring programmes must be designed on the basis of a conceptual understanding of the system where the level of refinement needed in a model (conceptual model) is proportionate to (i) the difficulty in making the assessments or predictions required, and (ii) the potential consequences of errors in those assessments (EC, 2003b). Hence, the more complex the system is, the more monitoring data are likely required to understand the system or to reach a chosen confidence level.

One could then ask: when is a monitoring programme sufficient for the WFD – does such a programme exist in Europe? This question can not be fully answered until the first WFD cycle has been completed and evaluated after 2015. The following example may, however, put some light on the issue. Larsen *et al.* (2002) conducted an inventory on the National Monitoring Programme for the Aquatic Environment in Denmark with a focus on statistical optimization. This was done parallel to a review carried out by the European Topic Centre on Water (ETC, 2002) where it was concluded that the Danish monitoring programme in general is sound and adequate in terms of number of stations, spatial coverage, parameters and frequencies. Even so Larsen *et al.* (2002) concluded that frequencies are far too low to prove various statistical hypotheses and that often several (>5) decades with measurements will be required to identify a given trend. As an example the number of yearly measurements (presently 10) at the existing stations in lakes aiming at estimating the average level of parameters such as nutrients and chosen biological indicators should be raised with factor 2–5 to get a relative precision[1]

[1]The relative precision describes the relative deviation of the 95% confidence interval to the geometric mean – thus the lower relative precision the more well-determined is the value.

of 20%. Another example is the groundwater monitoring stations, around 1000 wells in 70 monitoring areas that are typically sampled once per year. This sampling frequency only provides a precision level around 50% of a given parameter (nitrate), and raising this precision to about 20% requires 2–4 times more measurements. This indicates that even a rather intensive monitoring programme may not in all situations be sufficient to fulfil the requirements of the WFD within the relatively short time span until 2015.

At the Harmoni-CA WP4 workshops several participants stated that some data types required by the WFD are still not being collected. This is a challenge to be solved the nearest years (Arustiene, 2005; Demetriou, 2005; Crabtree, 2005; Håkansson, 2005; Jørgensen, 2005; Pataki, 2005; Purvina, 2005).

2.3. Data requirements for modelling

Often data requirements for a given model code are considered as if it is possible to give a general answer to the question: "*Are there enough data for use of this model code*?" Typically the indicative list of required data provided in connection with documentation and users' guides of model codes are considered as fundamental requirements also implying that the model will be able to provide useful results if such data are available. The model evaluation tool provided by the BMW project (http://www.rbm-toolbox.net/bmw/index.php) for instance includes the question: "*Are all the necessary data required for the implementation of the model available*?"

The opposite view that there is always enough data to set up and apply a model is advocated by some modellers. As an extreme example Refsgaard *et al.* (1999) established a catchment model for simulation of groundwater contamination from nitrate leaching alone from standard European level databases such as GISCO, EUROSTAT and EEA. Their view was that it is always possible to generate input data from secondary data sources through various transfer functions, and that it is important to assess the corresponding model prediction uncertainty (Thorsen *et al.*, 2001). According to this view the question "*Are there enough data*?" is ill posed. Instead the relevant question should be: "*Is the model prediction uncertainty small enough to be of any practical use in a water management context*?" This last question cannot be answered without considering the socio-economic context, i.e., how much is at stake economically and politically for stakeholders and decision makers, and it will therefore vary from case to case (Refsgaard and Henriksen, 2004; Højberg *et al.*, 2007).

Depending on the model approach the data requirements vary greatly both with respect to the need for time-series and system data such as soil maps and land use. The different model concepts provide different degree of insight in the physical system and have different requirements with respect to system data and data for model calibration and validation, as illustrated in Figure 2.1. While the empirical models only describe an observed relation between the variables of interest, the physical-based models can, theoretically, describe all state variables of the entire model domain, and thus hold the potential of providing the most detailed and correct representation of the physical system. Most distributed models are, however, a combination of distributed and lumped descriptions, where less important areas/processes are described by a lumped approach, which lowers the data requirements for the model application.

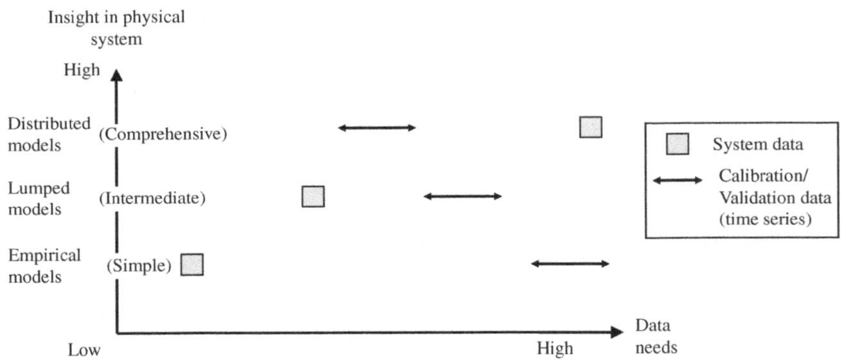

Figure 2.1 Illustration of the relation between data requirement and provided insight in the physical system for different model concepts

An implicit assumption in the use of models in praxis is that the model concept is correct, being either empirical, lumped or physically based, and the model parameters only need to be adjusted in order to mimic the site-specific system acceptable, i.e., it does not require a modification of the governing equation of the model code. If the model, on the other hand, is used to validate or refine the underlying relations/equations, a much more detailed description is generally required to study the relevant processes thoroughly, and much more data are required. This type of model use is not common in praxis but most often carried out in the research world by use of research models, where alternative formulations may be tested and compared to field observations from dense monitoring programmes designed for the specific purpose.

The data requirements for the different model types are listed in Table 2.1 as low/intermediate/high/very high and this is differentiated into system data and time series data in Figure 2.1. But how much data is that and when do you decide on selection of model type?

2.4. How and when in the modelling process is the necessary amount of data assessed?

The modelling process may, according to the HarmoniQuA project (Refsgaard *et al.*, 2005a; Scholten *et al.*, 2007, www.harmoniqua.org), be decomposed into five major steps grouping related tasks to perform at different stages in the model study, Figure 2.2. The contents of the five steps are:

STEP 1 (Model Study Plan). Some of the important issues of this step are to define the objectives and requirements for the model study. Questions to be addressed include: What is the purpose of the modelling – what are the questions we are searching an answer for? What are the consequences of being wrong? Where the latter question dictates the level of detail and precision required in the model study. What is the available data, and what are the needs of collecting additional data.

STEP 2 (Data and Conceptualisation). In this step the modeller collects all the relevant knowledge about the study basin and develops an overview of the processes and their interactions in order to conceptualise how the system should be modelled in sufficient detail to meet the requirements specified in the Model Study Plan.

STEP 3 (Model Set-up). With a conceptual model of the system the numerical model is constructed by transforming the conceptual model into a site-specific model that can be implemented in the model code.

STEP 4 (Calibration and Validation). In this step the models ability to reproduce the physical system is analysed. This is acquired first by calibration, where the model parameters are adjusted until the model response reproduces the reality within a user specified accuracy, and next by validation, where the model performance are compared to independent data not used for calibration.

STEP 5 (Simulation and Evaluation). In the final step the calibrated and validated model is used to carry out simulations to meet the objectives of the model study. These simulations may produce specific results that can be used in subsequent decision making (e.g., for planning or design purposes) or to improve understanding (e.g., of the hydrological/ ecological regime of the study area).

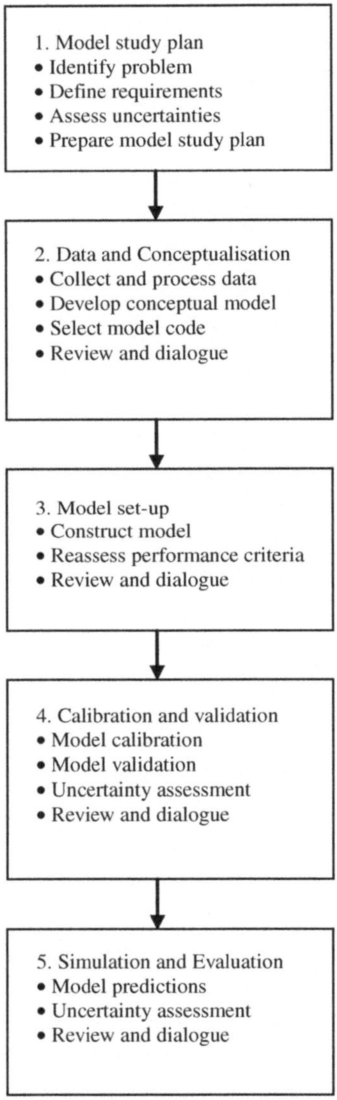

Figure 2.2 Decomposition of a model study into five major steps, simplified from Refsgaard *et al.* (2005a)

The flow chart in Figure 2.2 is a simplification. First of all each of the five steps are decomposed into several tasks. But maybe more importantly, it should be noted that the modelling process is not simply linear as depicted in Figure 2.2. Actually, one of the key characteristics of the modelling process described by Refsgaard *et al.* (2005a) and Scholten *et al.* (2007) is the many feedback processes and re-do loops caused both by model-technical reasons and by decisions made by the water manager after dialogues with stakeholders. Furthermore, it should be noted that the modelling process is interacting with the water management process. This is outlined in Refsgaard *et al.* (2007) and discussed in details in Bots *et al.* (2010).

In a model study data are required to conceptualise the system (STEP 2). Here the data requirements are mostly dependent on the complexity or heterogeneity of the physical system, but also on the detail and precision by which we need to know the physical system in order to provide the answers to the questions raised in the Model Study Plan within the specified precision. For simple and homogeneous system, or where the requirement to the precision is low, few data may be sufficient for an adequate understanding of the system. In highly complex and heterogeneous system, or where the consequences of being wrong are fatal and the required precision thus is high, a higher spatial and temporal resolution of the data are generally required. In STEP 4 data is needed for the calibration and validation of the model, where the requirements are mostly dependent on the performance criteria that have been defined for the model. If the criteria are low, it may be accepted that the model is only calibrated without a following validation. This lowers the requirements on the observations of the system response, but also lowers the confidence in the model results.

A first assessment of the data required to achieve a desired level of accuracy in model predictions have to be made already in STEP 1 and must be linked with the design of the WFD monitoring programme.

The amount of data required in a model study is thus controlled by the complexity of the physical system and the precision by which the model results are needed (model performance). The requirements can therefore not be expressed exact, like an absolute number of measurements per square unit, which is appropriate for all model studies. In contrary, the amount of data required varies from study to study depending on the objectives and requirements for the study. The question of how much data is required for modelling is therefore not well posed. Modelling is in principle possible with all levels of data quantity and quality. Instead it should be recognised that the quality and quantity of data will affect the uncertainty of the model outputs and as such the usefulness of the model simulations (Refsgaard *et al.*, 2005b). The relevant question is therefore what the requirements to the accuracy of the modelling are. Only when the answer to this question is given, the data requirements can be assessed in a meaningful manner.

2.5. The relation between data availability and model performance

In the scientific community there is no total agreement on the relationship between model complexity, data availability and predictive performance. One common illustration of the conceptual relationship is shown in Figure 2.3. It illustrates that the model performance in general increases with data availability, and that complex models are required to make full use of large data availability. It also illustrates, however, that for limited data availability the moderately complex models may often perform better than the very complex models. The reason for this is that the uncertainty of model predictions (the opposite of the predictive performance axis in Figure 2.3) consists of two terms: (a) the parameter uncertainty which increases with model complexity, because of an increasing number of parameters; and (b) the model structure uncertainty which decreases with model complexity. This implies that for complex models with many 'free' parameters that require data for calibration the increase in parameter uncertainty may be larger than the decrease in model structure uncertainty when we have exceeded a certain level of complexity.

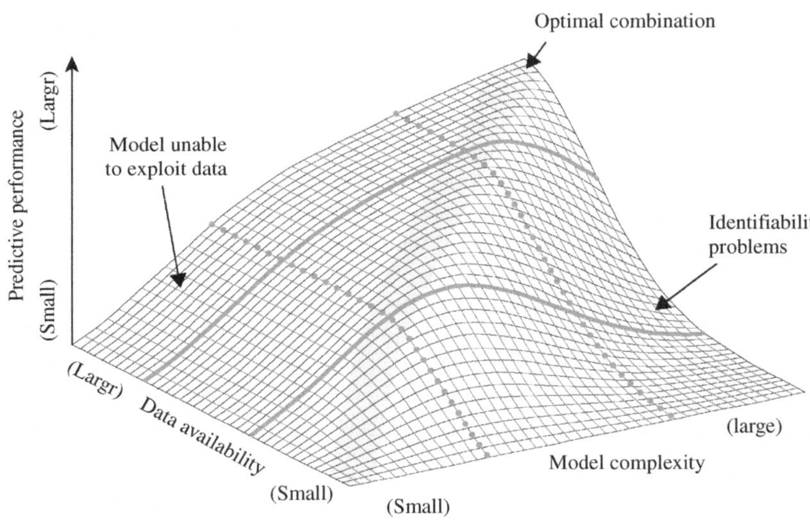

Figure 2.3 Schematic diagram of the relationship between model complexity, data availability and predictive performance (Modified from Grayson and Blöschl, 2000)

The view illustrated in Figure 2.3 originates from statistical theoretical considerations assuming that the number of free parameters requiring calibration increases with model complexity. This is, however, not always the case. For model codes with comprehensive application records there are often experience based recommendations for setting default parameter values, so that typically only a handful of parameters are adjusted by calibration in the particular catchment. In this way secondary data sources (all previous model applications in other locations) are used. This would correspond to moving left on the data availability axis towards larger data availability.

An often posed question is whether the data collected for monitoring purposes under the WFD are also sufficient for modelling. This issue is analysed by Højberg *et al.* (2007). They argue that because (a) the monitoring programme should be designed so that it provides data that are necessary to provide sufficiently reliable assessments; and (b) the water manager decides on the desired accuracy level of the analysis, including the use of models as required, the monitoring should in principle be sufficient also to support the necessary modelling studies. Otherwise the monitoring is not good enough. In practise this may, however, not always be the case, amongst others because the water manager in the first WFD cycle up to 2015 may not have been able to analyse thoroughly what is required to achieve a certain level of accuracy in model predictions at the time of designing the monitoring system. Furthermore, it is noted that this does often not fulfil the data needs of researchers for modelling analysis in relation to process studies, which are beyond the scope of WFD.

2.6. Conclusions on methodological considerations regarding data requirements

Data requirements depend on the system and model complexity and the purpose of the modelling study, which again depend on the socio-economic context, i.e., what is the problem, what are the conflicting interests and how much is at stake economically, socially and politically.

Therefore it is not possible to prescribe general data requirements for water management, and WFD, purposes. We will instead supplement the above general considerations by approaching this issue from the opposite side through reviewing the data availability in some European river basins and try to relate this to the types and character of the various water management problems and the types of modelling carried out in the respective basins.

3. DATA AVAILABILITY

3.1. Methodological considerations

The two main sources of information we have used to assess data availability in European river basins are:

- *Article 5 reports.* As an obligatory part of the WFD implementation EC member states have reported on the status of water bodies. These reports contain some information on data availability and modelling activities. These reports are available on the CIRCA website: http://forum.europa.eu.int/Public/irc/env/wfd/library?l=/framework_directive/implementation_documents_1/wfd_reports/member_states&vm=detailed&sb=Title).
- *The HarmoniRiB River Basin Network.* The HarmoniRiB EU FP5 RTD project collected data from eight representative river basins across Europe. The data availability and the actually collected data are described in reports available on the project website (http://workplace.wur.nl/harmonirib).

Several other EU funded research projects have collected river basin data that could potentially also be used to analyse data availability. This is for instance the case for several of the other CatchMod projects such as Euroharp (www.euroharp.org). One problem that makes use of information from these projects difficult is that they only collected data according to the needs of the research project, and this is only some of the available data. This implies that the water managers in practise have more data at their disposal than the data collected in the respective research projects. This would provide a biased view of data availability. We have therefore focused on the HarmoniRiB data that are more complete in this respect and made an additional effort to check that the information shown in the present report provides a correct picture of the full amount of data available to the water managers in these eight basins.

As mentioned earlier in this report WISE – the Water Information System for Europe – is under development by EC etc. to substitute a number of databases on environmental data on a European scale. This web site will, once filled with data, allow users to browse all information and data collected during Water Framework Directive implementation in Europe. A cartographic (map view) interface enables users to zoom into any specific area in Europe to get information on the status of specific water bodies. Here it will also be possible to find additional information, such as reports from Pilot River Basins and guidance documents covering large parts of water management. The aim is to develop a water information system with publicly accessible information relevant for water management on a European scale. We have not analysed the WISE data as an indicator of the data availability, because the data availability in the river basins are in all cases significantly larger than what is reported to the WISE database, either in terms of the amount of stations and/or the temporal resolution of the data, as WISE focuses on data collected to document compliance with the WFD.

3.2. Information from WFD Article 5 reports (Basis Analysis)

EC member states have recently reported on the status of the water bodies – the so-called Article 5 reports (see link above). These reports are rather heterogeneous with respect to style and length. A quick survey of some of the country reports available in English (Cyprus, Bulgaria, Denmark, Estonia, Greece, Ireland, Latvia, Malta and Romania) show that most of the reports emphasise the following with respect to data availability:

- Data are still insufficient for a proper characterisation of the status of many water bodies; most often in terms of biological and ecological data on surface waters but also priority and hazardous substances as well as surveys of contributions of nutrients from diffuse pollution, both in surface water and groundwater, are mentioned.
- There is a need for further data collection to fulfil the WFD requirements and this will be done in the coming years.
- Much of the characterisation and classification has been carried out based on expert knowledge.

Most Article 5 reports do not directly address the current data availability and accessibility in their reports and most do not address the need for modelling nor the question to which extent the existing data are sufficient for modelling purposes. However, some reports mention modelling projects already carried out, which has helped in the characterisation etc. of water bodies and a few mentions modelling as a possible tool to overcome some of the future challenges and gaps in data etc.

It appears from these reports that the national authorities responsible for WFD implementation in many cases feel that the existing data are not sufficient for all the WFD requirements, and that they therefore foresee the need to improve the data collection during the coming years. The language in these reports is, however, not very rigorous, and they do not explicitly address the need of data for modelling. The conclusion we can draw should therefore be taken with some reservations.

At this early stage of the WFD implementation it is quite understandable that member states are cautious regarding data adequacy. Partly, because it is impossible at this first WFD iteration to know what the optimal data amount is to design the programme of measures that will be decided in a participatory decision process and to efficiently monitor the effects its implementation. And partly, because data adequacy is not something static. External pressures are dynamic, and often increasing, and therefore the data needs will also often increase correspondingly.

3.3. HarmoniRiB River Basin Network

Within the EU funded HarmoniRiB project WFD relevant data from a network of eight river basins throughout Europe have been collected and analysed (Refsgaard *et al.*, 2005b, 2006). These basins have been selected to ensure a good coverage across Europe in terms of eco-regions, types of water problems, socio-economic conflicts and amount and quality of existing data. In addition, two of the river basins (Odense and Jucar) are also included in the Pilot River Basin Network, where the EC guidance documents have been tested. The location and characteristics of the eight basins are shown in Figure 3.1 and Table 3.1. Brief descriptions of each of the basins and their respective water resources problems together with a brief description of the data availability and accessibility are provided in the Appendix. It should be noted that the basins are not research basins, and that the types of problems vary significantly from predominantly agriculturally related water quality in northern Europe to water shortage in southern Europe and to some basins experiencing severe floods.

The data availability may be characterised as representative for its region and type of water management problems with some minor reservations. First of all, these basins were selected for the research project amongst others because the basin organisations were considered to be relatively resourceful as compared to similar organisations in their respective countries/regions. Secondly one should be somewhat cautious to claim representativeness from such a small sample.

It is not feasible and it is outside the scope of this Guidance Report to list all the data that are available for the selected river basins. Instead some variables/data types have been selected. They serve as indicators of data availability and may not be fully representative in all cases.

A summary of the data availability is given in Table 3.2. It appears from Table 3.2 that there are large differences in the amount of data in the various basins. It is also clear that the data collection has been tailored to the water resources problems in the basins. For instance the river basins with the largest drought/water shortage problems (Geropotamou-GR, Candelaro-IT and Jucar-ES) are in the top of data coverage with respect to meteorological data. On the other hand, the amount of surface water quality data is much higher in countries like Germany, Netherlands and Denmark where eutrophication is a key WFD problem compared to the Mediterranean basins in Greece, Southern Italy and Spain, where non-sustainable groundwater abstractions is the key problem. Similarly, much more data on groundwater quality exist in the Danish river basin, where groundwater quality is the key problem related to drinking water problem, than in other countries where this is not so major an issue.

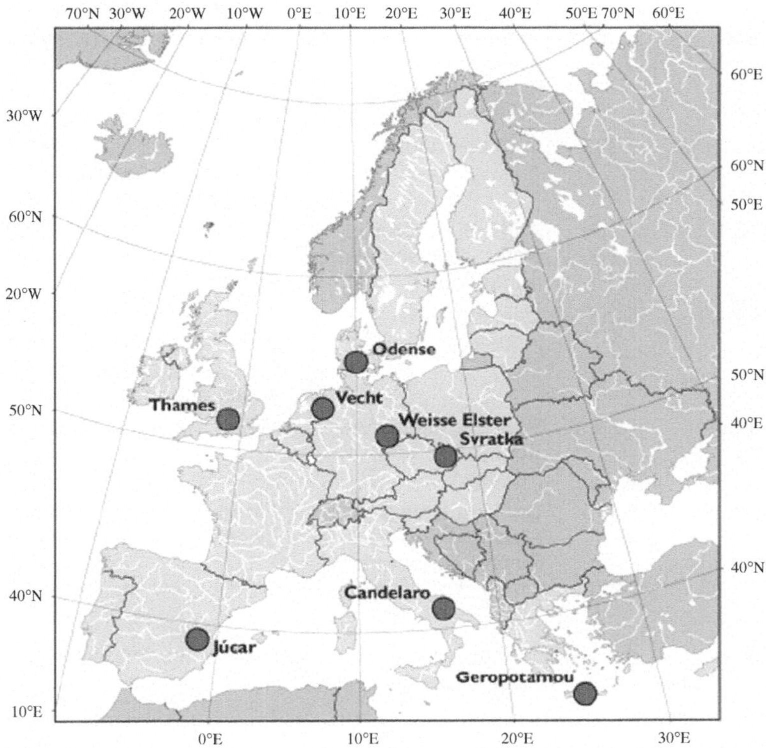

Figure 3.1 Location of the HarmoniRiB network of representative river basins

Table 3.1 Key characteristics of the HarmoniRiB network of representative river basins

Country River Basin	Area (km^2)	Main water uses	Main conflicting interest
DK, Odense	1,046	Public water supply, recreation, nature	Agricultural contamination; groundwater abstraction depletes stream flow and wetlands
GR, Geropotamou	600	Irrigation, tourist	Water shortage, water quality, oversized dam, salt intrusion, difficulties in sharing water among municipalities
ES, Jucar	42,989	Irrigation, hydroelectric, tourist supply, industry	Water shortage, farming use; hydroelectrical use; tourist water demand
IT, Candelaro	2,330	Irrigation, industry	Water shortage; rainfall rates decrease; intensive horticultural farming.
NL+DE Vecht	3,780 (1980 in NL)	Agriculture, drinking water, receiving water, recreation	Agriculture, water quality, ecology, flooding – room for water retention
DE, Weisse Elster	5,200	Drinking water, industry	Point and non-point sources; wastewater and contaminated sites; strong economic and social changes.
UK, Thames	12,920	Public water supply, Ecosystem, recreation	Water supply *vs.* ecology
CZ, Svratka	4,115	Drinking water, electrical power, recreation, nature	Flood protection, minimum discharges, water quality

Examples of the models that have been supported by the available data are given in Table 3.3. By comparing these tables with Table 3.2 it appears that modelling has been carried out in domains with perceived water resources problems. Altogether this gives, not surprisingly, a clear correlation between (a) water resources problem domains; (b) data availability; and (c) previous modelling studies.

Table 3.2 Data availability for selected data types in the eight HarmoniRiB river basins

Domain – data type	Odense 1,046 km^2		Geropotamou 600 km^2		Jucar 42,989 km^2		Candelaro 2,330 km^2		Vecht (Dutch part) 1,980km^2		Weisse Elster 5,200 km^2		Thames 12,920 km^2		Svratka 4,115 km^2	
	Stations	Km^2/station	Stations	Km^2/station	Stations	Km^2/station	Stations	Km^2/station	Stations	Km^2/station	Stations	Km^2/station	Stations	Km^2/station	Stations	Km^2/station
Water resources assessment																
• Daily precipitation	7	149	11	55	634	67	27	86	12	165	61	85	199	65	15	274
• River discharge	10	105	3	200	138	312	9	259	6	330	18	289	160	81	17	242
• Groundwater level	60	17	18	33	1837	23	37	63	1200	1.7	130	40	35	369	12	343
Surface water quality, ecosystems																
• Physio-chemical (daily or better)	4	262	0	∞	13	3307	0	∞	8	473	2	2600	10	1292	0	∞
• Physio-chemical (weekly or less)	2	523	0	0	232	185	6	388	1170	1.7	50	104	2500	5	15	274
• Biological	29	36	0	∞	251	171	0	∞	1699	1.2	20	260	43	300	12	343
Agriculture, non-point sources			Note #1		Note #1		Note #1						Note #1		Note #1	
• Soil profiles or maps	3619								1:2.5 10^5		1:10^6 1:200 000					
• Management (crops, yield, fertilizer, pesticide application, irrigation etc.)	Farm level								Various		County level					
Groundwater quality																
• Physico-chemical	1725	0.4	14	43	544	79	37	63	480	4	30	173	500	26	5	823
• Specific synthetic pollutants	300	3.5	0	∞	Note #1		0	∞	480	4	0	∞	Note #1		0	∞
• Specific non-synthetic pollutants	300	3.5	0	∞	Note #1		0	∞	480	4	0	∞	Note #1		0	∞

Note #1: No information available

Table 3.3 Examples of models supported by data from the eight HarmoniRiB river basins

Domain	Model complexity	Odense, DK	Geropotamou, GR	Jucar, ES	Candelaro, IT	Vecht (Dutch part), NL	Weisse Elster; GE	Thames; UK	Svratka; CZ
Water resources assessment									
	GIS tools	GIS tools		GIS tools	GIS tools	GIS tools	GIS tools		
	Intermediate – Planning	Groundwater availability. Regional models. MIKE SHE	Water availability. Sacramento, MODFLOW	Rainfall-runoff modelling	Water availability assessment. Owner made model.	MODFLOW/ MLAEM- NAGROM SWAP	WASIM-ETH	IHACRES LOW FLOWS 2000	Water management studies
	Comprehensive – Design	Streamflow depletion due to groundwater abstraction. Local models MIKE SHE	Water availability. Sacramento, MODFLOW			MODLFOE/ MLAEM NL-CAT MIPWA	WASIM-ETH		
	Research – Process studies					NL-CAT Many field scale models. Various GGOR supporting models			
Flooding									
	Simple – Screening:				Location of areas subjected to flooding. GIS tools				
	Intermediate – Planning				SWAT	SOBEK	HEC-RAS 1D	PDM rainfall-runoff. KW channel flow routing	
	Comprehensive – Design					SOBEK (and others)		WINFAP-FEH; FEH; ReFH	Flood risk assessment. MIKE 11

(*continued*)

Table 3.3 *Continued*

Domain	Model complexity	Odense, DK	Geropotamou, GR	Jucar, ES	Candelaro, IT	Vecht (Dutch part), NL	Weisse Elster; GE	Thames; UK	Svratka; CZ
	Research – Process studies								Regional studies of flood protection measures
Surface water quality, ecosystem									
	Simple – Screening:	GIS tools			GIS tools	GIS tools, Excel (Rebelgroup)	MONERIS		
	Intermediate – Planning			Hydro ecological modelling	SWAT, QUAL2E, REMM. Buffer strip effectiveness.	WFD-explorer (in development). Balance model de Straat		QUESTOR; QUASAR; STARBUGS; RIVPACS	Planning of wastewater treatment.
	Comprehensive – Design	Odense Fjord water quality. MIKE 3				NL-CAT	WASP5. ATV-Model with additional measurements		
	Research – Process studies					NL-CAT			Project "Clean Svratka river basin" – removal of blue green algae from Brno reservoir
Agriculture, non-point sources									
	Simple – Screening:	GIS tools Effects of management scenarios on P load.			GIS tools. Effects of management scenarios on N and P load.		GIS tools EXPORT Coefficient Models, R USLE		Nutrient load. Regression models.

(*continued*)

Table 3.3 *Continued*

Domain	Model complexity	Odense, DK	Geropotamou, GR	Jucar, ES	Candelaro, IT	Vecht (Dutch part), NL	Weisse Elster; GE	Thames; UK	Svratka; CZ
	Intermediate – Planning	N leaching. NLES, Daisy			GLEAMS, Index methods and metamodeling. Critical area selection		HBV-P/N; AGNPS EROSION 3D (in subcatchments) INCA		
	Comprehensive – Design	Effects of management scenarios on N leaching. Daisy			SWAT. Basin management plans		SWAT		
	Research – Process studies	Euroharp + HarmoniRiB + GEUS and NERI research (Daisy +P-index)			New modules development for SWAT. For intermittent rivers.				
Groundwater quality									
	Simple – Screening:	GIS tools			GIS tools		GIS tools Meta-CANDY		
	Intermediate – Planning	Groundwater vulnerability. Local models. MIKE SHE, MODFLOW		MODFLOW	Impact of agricultural practices on groundwater quality. GLEAMS		Soil water models like CANDY		
	Comprehensive – Design								
	Research – Process studies	Groundwater age MIKE SHE. Nitrate reduction in groundwater. Daisy/MIKE SHE. Additional data collection							

By combining the information on data availability for indicator variables (Table 3.2) with the information on previous modelling activities in the respective river basins (Table 3.3) we have prepared Table 3.4 showing an indication of the suitability of the existing data for various levels of model complexity and model applicability. There is an element of subjective assessments in this. However, as the data availability and the past modelling studies as previously noted are closely correlated, we believe that the picture depicted by Table 3.4 is a reasonable reflection of the real differences between the river basins.

3.4. Discussion and conclusions regarding data availability in river basins

The analyses of the Article 5 reports and the HarmoniRiB basins lead to two conclusions with some apparent differences. On the one hand, it appeared from the Article 5 reports that data availability in many cases may not be sufficient for the entire WFD implementation. On the other hand, we concluded that the data availability in the HarmoniRiB basins was sufficient for some level of modelling.

An interesting question is then if this level of model complexity/applicability indicated in Table 3.4 is sufficient for WFD implementation. This cannot be answered directly. On one hand it appears from comparison of Table 3.4 and river basin descriptions in the Appendix that there is a high degree of correlation between the domains of water resources problems and the number of stars in Table 3.4 suggesting that the data availability may be sufficient for the first iteration of the WFD implementation. However, on the other hand it is not possible at this early stage of the WFD process to tell exactly what the requirements to accuracy of model predictions are. This will depend on the economic assessments of the programme of measures and, not least, on the stakeholder views as they appear through the public participation process.

The message from the Article 5 reports, namely that the data in many cases may not be sufficient for all aspects of the coming WFD implementation, therefore suggests that the level of modelling required to support the WFD implementation will be somewhat higher than what can be performed with the present data availability. The important point here is that the requirements to data from a socio-economic (water management) point of view is consistent with the data requirement to modelling, simply because it is the socio-economic requirements that determine what kind of modelling should be performed.

Table 3.4 Data suitability to support model studies at varying complexity in the eight HarmoniRiB river basins

Model complexity	Odense, DK	Geropotamou, GR	Jucar, ES	Candelaro, IT	Vecht, NL	Weisse Elster, DE	Thames, UK	Svratka, CZ
Water resources assessment	***	***	***	***	***	***	***	***
Flooding	*	*	*	*	**	**	**	***
Surface water quality, ecosystem	***	*	**	*	***	***	***	**
Agriculture, non-point pollution	***	*	**	**	***	***	*	*
Groundwater quality	***	*	**	**	***	*	**	*

Level of model complexity/application:
* Simple – screening
** Intermediate – planning
*** Comprehensive – design
**** Research – process studies

4. DATA ACCESSIBILITY

4.1. Accessibility for water resources management

Participants in the workshops organised under the umbrella of Harmoni-CA WP4 pointed out several times that it is often a problem to obtain data collected by other institutions (Steenstra *et al.*, 2004; Kamphorst *et al.*, 2005; Arustiene *et al.*, 2005; Jørgensen *et al.*, 2006). Consequently, one of the main recommendations from all four workshops to policy makers was to secure that all environmental data are freely available in practise, i.e., to ensure centralised data collection and centralised databases with free access – or at least available at low costs, as too high costs are a serious constraint for many organisations (Kamphorst *et al.*, 2005).

In connection with two of the workshops ("Remote Sensing and Data Assimilation Techniques" (Steenstra *et al.*, 2004) and "State-of-the-art on Existing Monitoring Programmes around Europe" (Kamphorst *et al.*, 2005)) participants described existing monitoring programmes on a regional or national basis. Many mentioned the problem of getting transboundary data or even just to co-operate across borders or obtain cross-border data harmonisation (Steenstra *et al.*, 2004). Another problem relates to getting data at the right level to satisfy the needs of the public in stakeholder involvement procedures (Steenstra *et al.*, 2004). Also the data format was mentioned as a problem: data are not always available digitally (Steenstra *et al.*, 2004; Demetriou, 2005; Zupan, 2005). Even different institutes collecting the same data without co-operation was mentioned (Steenstra *et al.*, 2004). Closely related data stored in different databases at different institutes or even in private companies (Pataki, 2005; Koussis, 2005; Heller, 2005; Demetriou, 2005) were also pointed out as bottlenecks in data access. It was also mentioned that in spite of politically stated good intentions, some databases are in practise not accessible from outside the monitoring authorities (Koussis, 2005; Heller, 2005).

In summary, the five main constraints to data accessibility are:

- Economical. Most data are publicly available free of costs or at a nominal cost corresponding to the work of making them available in practise such as the cost of copying. For some data in some countries the costs are significant with the result that the data are used less than they would have been otherwise. This is for instance often the case for meteorological data and topographical data.
- Political. Sometimes existing data are not openly available to all interested parties. This can for instance be the case for data collected by a stakeholder who does not want other organisations, with conflicting interests, to have access to the data.

- Data formats. Data are more and more made available in digital formats, but still some data are only available on paper and often the formats of different data types are very different and the level of their documentation so lacking that it provides practical constraints for data use.
- Transboundary co-operation. It is often difficult to get data from organisations in other countries (even within the same river basin) and these data are often not harmonised with respect to formats and documentation.
- Fractionised databases. Data often exist in many databases with different data owners each having their own data formats and dissemination policies. This fractioning of data makes it difficult in practise to get an overview of the data availability and requires a lot of work for case specific data processing. Although considerable efforts are made in recent years to homogenise and integrate databases the fractioning still poses a major constraint in practise.

Internationally there are declarations of intent prescribing free assess to data – e.g., the Aarhus Convention[2] (UNECE, 1998) that grants the public rights regarding access to information and public participation as well as access to justice. It focuses on interactions between the public and public authorities. Article 4 of the convention secures the public access to environmental information from public authorities and, in some cases, from private parties, within a reasonable amount of time. Further, Article 5 imposes Parties and public authorities to collect and disseminate environmental information. Paragraph 3 and 4 in this article specifically mentions "Electronic databases" containing "Environmental information" as well as "National state-of-the-environment reports" within information on quality and pressures on the environment (UN, 2000).

A number of European initiatives have been launched during the last years to implement the Aarhus Convention, e.g., the INSPIRE project and the WISE web-portal.

The INSPIRE "Infrastructure for Spatial Information in Europe" is an EU initiated co-operative project between DG Environment, EUROSTAT and the

[2]The UNECE (United Nations Economic Commission for Europe) Convention on Access to Information, Public Participation in Decision-making and Access to Justice in Environmental Matters, usually known as the Aarhus Convention, was signed on June 25, 1998 in the Danish city of Aarhus. It entered into force on 30 October 2001. As of November 2005, it has been signed by 40 (primarily European) countries and ratified by 37. It has also been ratified by the European Union, which has begun applying Aarhus-type principles in its legislation, especially the Water Framework Directive.

Joint Research Centre. The aim has been to trigger the creation of a European spatial information infrastructure that delivers integrated spatial information services to all users. These services should allow the user to identify and access spatial or geographical information from a wide range of sources, from the local level to the global level, in an inter-operable way for a variety of uses. The target users of INSPIRE include policy makers, planners and managers at European, national and local level and the citizens and their organisations. Possible services are the visualisation of information layers, overlay of information from different sources, spatial and temporal analysis, etc. The initiative has produced an INSPIRE directive (EC, 2007) on establishing an Infrastructure for Spatial Information for the European Community.

WISE – the Water Information System for Europe – is another co-operation project involving European environmental institutions as the European Commission (in particular DG Environment, Joint Research Centre and EUROSTAT), the European Environmental Agency and the Member States in modernising the collection and dissemination of information on water policy etc. across Europe. The output is a web portal launched at the World Water Day on 22 March 2007. The web portal provides access to data and reports related to the Water Framework Directive, the Urban Waste Water Treatment Directive, the European Environment Information and Observation Network on Water (EIONET/Water).

The WISE-RTD web portal "Consolidated Experiences in Water Management" is a portal linked to WISE, developed by Harmoni-CA. This web-portal is a gateway allowing users to get access to information on water-related (research) projects with relevance to the WFD.

INSPIRE and WISE will, when fully implemented, provide easy public access to many of the WFD relevant data and in this way some of the above problems will be solved. There are, however a number of additional data that are also required and used by water managers that will not be included in the INSPIRE and WISE. Attention is required to ensure easy accessibility for all relevant stakeholders to these data.

4.2. Accessibility for research purposes

Data collection is often a major task in research projects. In projects where existing data are used the data collection is often cumbersome and requires a lot of resources, because the data access is difficult with many practical and economical constraints. In projects that collect new field data with standard or new experimental equipment the data collection is usually very costly. Often data collected in one research project is not used in many other projects due to lack of proper data documentation and dissemination after the termination of old projects. The same data are therefore often collected several times by different research projects. This is obviously non-optimal and requires a lot of research resources both in terms of costs and manpower that could have been utilised much better.

Therefore, there is a need to make river basin data easily accessible for the scientific community. In this respect data for two different types of research projects are relevant:

- *Water resources management.* These projects are typically multi-disciplinary research studies linking hydrology, ecology and socio-economic aspects, e.g., related to the WFD implementation. Here existing 'real-life' data are relevant. An example of such data is the data from the HarmoniRiB river basin network that have been collected with the aim of making them freely available for non-commercial research purposes.
- *Hydrological process studies.* New data from experimental river basins with new sophisticated field equipment aimed at providing data that can enhance the understanding of the physical processes. An example of such data are those that will be provided through the new Hydrological Observatories to be established under FP7 under the Capacities/Research Infrastructures programme.

5. DATA QUALITY – UNCERTAINTY

In addition to the availability and the accessibility of the data, the quality of the data is crucial. If data are not trustworthy they are not useful. Some data are known to be of good quality with well documented measurement, analysis and quality control procedures. Some data are perceived to be of questionable quality and other data even known to be of poor quality. There are many possible reasons for poor or questionable data quality, such as:

- Field or laboratory work may have been carried out with improper equipment.
- Field or laboratory work may have been carried out by use of inadequate procedures.
- Data analysis may be inadequate.
- Quality control may be lacking or inadequate.

- The available documentation of data collection and processing may be lacking or inadequate.
- The representativeness (in time and space) of the data for use towards solving a particular problem may be poor due to inadequate station network and/or sampling frequency.

Data quality should not be characterised categorically just as good or poor with the implicit implication that the data are useful or useless, respectively. One reason for this is that the scale of data quality should be more differentiated and for instance expressed as data uncertainty in terms of probability distributions. Furthermore, the usefulness of uncertain data depends on the context in which they are used. Data with some level of uncertainty may be useful for a first screening of a problem, but may be inadequate for a comprehensive analysis with high demands to predictive accuracy.

Data quality can be characterised in terms of data uncertainty. Uncertainty is here defined as the degree of confidence that a person has in the data and the probabilities of the data (Klauer and Brown, 2003). It is noted that by adopting this definition uncertainty is considered as subjective implying that we can not know what the true uncertainty is, but different persons can come to consensus about their respective uncertainty assessments.

Understanding the uncertainty in environmental data and systems is essential for making robust and wise water management decisions (Funtowicz and Ravetz, 1990; Pahl-Wostl, 2002). Therefore characterisation of data uncertainty is becoming an important basis for modern water resources management. Concepts and tools for characterisation and handling of data uncertainty are described in Refsgaard *et al.* (2005b, 2006) and Brown *et al.* (2006). Literature surveys were made to support the assessment of uncertainty in different types of data (Van Loon and Refsgaard, 2005).

In environmental modelling studies field data usually have a spatial and temporal scale of support that is different from the one at which models operate. This calls for a methodology for rescaling data uncertainty from one support scale to another (Heuvelink and Pebesma, 1999; Brown *et al.*, 2006).

The practical implications of this are that data uncertainty should be assessed and that information on data uncertainty should be stored in as an important element of the data documentation. A problem in this respect is that standard databases today are not designed to enable storage of data uncertainty (Refsgaard *et al.*, 2006).

6. CONCLUSIONS AND RECOMMENDATIONS

6.1. Data availability

Conclusions

- The data availability of European river basins is generally much larger than the data reported to the Water and Information System for Europe (WISE). WISE will give access to data that are reported compulsory to document WFD compliance. However, much more data exist, and local water managers will usually rely on their own databases containing the full amount of existing and accessible data.
- There is a large diversity of data availability in river basins across Europe. For some data types like meteorological and river discharge data the data coverage only varies moderately among river basins. When it comes to data on surface water quality and biology and on groundwater quality there are, however, very large differences. This difference appears to a large extent to reflect the seriousness of the pressures (population density, pollution, etc.); however, the water management traditions and the socio-economic policy priorities may also be important factors.
- The level of complexity of modelling carried out in river basins appears in general to be proportional to the amount of data available. This may be because both the amount of data and the need for modelling to a large extent is driven by the level of pressure on the water resources, i.e., on the perceived water management problems. So maybe, at least in areas and domains with severe problems, the interaction between data and modelling is in fact larger than we think – even if it many times may not be explicitly recognised by all involved actors.
- Different levels of modelling complexity require different amounts of data, but it is not possible in general to answer the question on how much data is required for carrying out a modelling study. In general, it is always possible to establish and run a model. If primary data do not exist secondary data and transfer functions can be used. The relevant question, however, is if the model results with the given data availability can produce predictions that are sufficiently reliable to be of practical use. The question of data requirements is therefore linked to the accuracy requirements which again vary from case to case depending on the socio-economic context.
- The analysis of data requirements for modelling for supporting the WFD implementation should start early in the WFD process, namely in connection with design of the monitoring network. However, it is not possible to

complete this analysis until after the stakeholders views to the programme of measures are known through the public participation. At that time it will in some cases be too late to start and to make use of additional data collection. Lack of adequate data for the first round of the WFD cycle in 2015 will therefore have to be taken into account during the next six-year WFD cycle.
- Analysis of Article 5 reports suggests that data availability in many cases may not be sufficient for all aspects of the coming WFD implementation. This implies that the level of modelling required to support the WFD implementation will be somewhat higher than what can be performed with the present data availability. So, although the present data availability appears to be adequate for some level of modelling, it may in many cases not be sufficient for the level of modelling required from a socio-economic (water management/WFD) point of view.

Recommendations

- More readily available data than presently reported to WISEI are required to enable pan-European scientifically based analyses of WFD related water resources issues. Much of this data exist locally and are used for analyses in the river basins.
- The analysis of data requirements for WFD implementation in a river basin must consider also the data needs for the modelling studies that are expected to be required to support the water resources management. This should be included already at the time of designing the monitoring network.

6.2. Data accessibility

Conclusions

- The principles for data access as outlined in the Aarhus convention and the EU follow-up initiatives such as INSPIRE and WISE are sound and aim at providing easy public access to data. Thus there is no doubt that there is an underlying trend towards easier access to data. In spite of this there are still large problems with data access in practise. Some of the key problems are related to:
  - Economical. For some data in some countries the costs are significant with the result that the data are used less than they would have been otherwise. This is for instance often the case for meteorological data and topographical data.

  - Political. Sometimes existing data are not openly available to all interested parties. This can for instance be the case for data collected by a stakeholder who does not want other organisations, with conflicting interests, to have access to the data.
  - Data formats. Some data are only available on paper, often the formats of different data types are very different and the level of their documentation so lacking that it provides practical constraints for data use.
  - Transboundary co-operation. It is often difficult to get data from organisations in other countries (even within the same river basin) and these data are often not harmonised with respect to formats and documentation.
  - Fractionised databases. Data often exist in many databases with different data owners each having their own data formats and dissemination policies. This fractioning of data makes it difficult in practise to get an overview of the data availability and requires a lot of work for case specific data processing.
- When the new EU data initiatives INSPIRE and WISE are fully implemented in some years some of the above data accessibility problems will be solved, at least with respect to the data for which INSPIRE and WISE will provide access. However, many data that are used for water management at local and river basin scale will not be accessible through INSPIRE and WISE, which focus on data with some level of aggregation and on data legally required to document WFD compliance. The remaining data that are vital for detailed analysis, such as modelling, may still have some of the accessibility problems outlined above.

Recommendations

- Efforts should be made to ensure data accessibility in practice for all data required to do detailed analysis at local and river basin scale – also for data that are not reported to the INSPIRE and WISE web portals but are stored only in various national or regional databases.

6.3. Data quality – uncertainty

Conclusions

- In addition to the availability and the accessibility of the data, the quality of the data is crucial. If data are not trustworthy they are not useful. Understanding the uncertainty in environmental data and systems is essential for making robust and wise water management decisions.

Recommendations

- Data quality should be characterised in terms of uncertainty. This uncertainty information should be stored as part of the data documentation.

6.4. Data for research purposes

Conclusions

- Data requirement for research purposes vary tremendously depending on the objective of the research study. As the nature of research is to gain new knowledge, there will seldom be sufficient existing data on the particular research topic. If for instance the objective is to improve the understanding of flow and reactive transport processes at a certain scale in a given area then it will most often be required to design a dedicated field monitoring programme just for the research project. If, on the other hand the research is focussing on water management issues and decision making there will often be sufficient data on the hydrological system, but instead it may be required to collect information on e.g., stakeholder views and behaviour.
- Collection of existing data for research projects is often cumbersome and requires a lot of resources, because the data access is difficult with considerable practical and economical constraints.

Recommendations

- The research community should have easy access to datasets from representative river basins for conducting WFD related research studies. Data collected in selected EC funded research projects should be documented and uploaded to a pan-European database and afterwards be made available for future research projects.
- There is a need for the research community to have easy access to new data from experimental river basins with new sophisticated field equipment aimed at providing data that can enhance the understanding of the physical processes. An example of such data are those that will be provided through the new Hydrological Observatories that are planned to be established under FP7 through a 2008 call for proposals under the Capacities/Research Infrastructures programme.

7. REFERENCES

Arustiene, J. (2005). Groundwater monitoring in Lithuania. In: Kamphorst, E., Jørgensen, L.F., van Griensven, A. and Vanrolleghem, P.A. (Eds.) (2005). *Joint Use of Modelling and Monitoring for Implementing the Water Framework Directive.* Workshop report, 1st workshop: State of the art on existing monitoring programmes around Europe. Harmoni-CA. http://www.harmoni-ca.info/

Arustiene, J., Vaitiekuniene, J. and Jørgensen, L.F. (Eds.) (2005). *Joint Use of Modelling and Monitoring for Implementing the Water Framework Directive.* Workshop report. Harmoni-CA. http://www.harmoni-ca.info/

Beven, K. (1996). A discussion of distributed hydrological modelling. In: Abbott, M.B., Refsgaard, J.C. (Eds) *Distributed Hydrological Modelling.* Kluwer Academic Publishers, 255–278.

Blind, M. and de Blois, C. (2003). *The Water Framework Directive and its Guidance Documents – Review of data aspects.* Chapter 5 in Refsgaard, J.C. and Nilsson, B. (Eds.) Requirements Report. Geological Survey of Denmark and Greenland, Copenhagen. http://workplace.wur.nl/harmonirib.

Bots, P., Gooch, G., McIntosh, B.S. and Pahl-Wostl, C. (2010). Understanding the role of perception and valuation in the development and use of models for water management. In: Vanrolleghem, P.A. (Ed.) *Integrated Assessment for WFD implementation: Data, Economic and Human Dimension – Water Framework Directive Series, Volume 2.* IWA Publishing, London, UK, (this volume).

Brown, J.D., Heuvelink, G.B.M. and Refsgaard, J.C. (2005). An integrated framework for assessing and recording uncertainties about environmental data. *Water Science and Technology* **52**(6), 153–160.

Brown, J.D. and Heuvelink, G.B.M. (2006). *Data Uncertainty Engine (DUE) User's Manual.* University of Amsterdam. http://workplace.wur.nl/harmonirib.

Crabtree, B. (2005). River quality monitoring for WFD implementation in England and Wales. In: Kamphorst, E., Jørgensen, L.F., van Griensven, A. and Vanrolleghem, P.A. (Eds.) *Joint Use of Modelling and Monitoring for Implementing the Water Framework Directive.* Workshop report, 1st workshop: State of the art on existing monitoring programmes around Europe. Harmoni-CA. http://www.harmoni-ca.info/

Demetriou, C. (2005). The surface and groundwater monitoring network in Cyprus. In: Kamphorst, E., Jørgensen, L.F., van Griensven, A. and Vanrolleghem, P.A. (Eds.) *Joint Use of Modelling and Monitoring for Implementing the Water Framework Directive.* Workshop report, 1st workshop: State of the art on existing monitoring programmes around Europe. Harmoni-CA. http://www.harmoni-ca.info/

EC (2000). Directive 2000/60/EC of the European Parliament and of the Council of October 23 2000 establishing a framework for Community action in the field of water policy. *Official Journal of the European Communities, L327/1–L327/72, 22.12.2000.*

EC (2003a). *Water Framework Directive, Common Implementation Strategy, Guidance Document No. 1, IMPRESS, Analysis of Pressures and Impact.* http://forum.europa.eu.int/Public/irc/env/wfd/library

EC (2003b). *Water Framework Directive, Common Implementation Strategy, Guidance Document No. 7, Monitoring.* http://forum.europa.eu.int/Public/irc/env/wfd/library

EC (2003c). *Water Framework Directive, Common Implementation Strategy, Guidance Document No. 9, Planning processes.* http://forum.europa.eu.int/Public/irc/env/wfd/library

EC (2006). Directive 2006/118/EC of the European Parliament and of the Council of 12 December 2006 on the protection of Groundwater against pollution and deterioration. *Official Journal of the European Communities*, L 372/19 – L372/31, 27.12.2007

EC (2007). Directive 2007/2/EC of the European Parliament and of the Council of 14 March 2007 establishing an Infrastructure for Spatial Information in the European Community (INSPIRE). *Official Journal of the European Communities*, L108/1 – L108/14, 25.4.2007

ETC WTR (2002). *Evaluation of NOVA-2003*. 90 pp. European Topic Center on Water & European Environmental Agency. http://www.dmu.dk/NR/rdonlyres/085E220D-16CF-4916-B738-D68C5F30DFA9/0/Evaluation_NOVA2003_Final_Report.pdf

Funtowicz, S.O. and Ravetz, J.R. (1990). *Uncertainty and Quality in Science for Policy.* Kluwer, Dordrecht, The Netherlands.

Grayson, R. and Blöschl, G. (2000). Spatial modelling of catchment dynamics. In: Grayson and Blöschl (Eds.) *Spatial Patterns in Catchment Hydrology: Observations and Modelling*. Chapter 3. Cambridge University Press, UK, pp. 51–81.

Heller, J. (2005). Surface water monitoring in Czech Republic. In: Kamphorst, E., Jørgensen, L.F., van Griensven, A. and Vanrolleghem, P.A. (Eds.) *Joint Use of Modelling and Monitoring for Implementing the Water Framework Directive.* Workshop report, 1st workshop: State of the art on existing monitoring programmes around Europe. Harmoni-CA. http://www.harmoni-ca.info/

Heuvelink, G.B.M. and Pebesma, E.J. (1999). Spatial aggregation and soil process modelling. *Geoderma* **89**, 47–65.

Højberg, A.L., Refsgaard, J.C., van Geer, F., Jørgensen, L.F. and Zsuffa, I. (2007). Use of Models to Support the Monitoring Requirements in the Water Framework Directive. *Water Resources Management.* http://dx.doi.org/10.1007/s11269-006-9119-y

Håkansson, B. (2005). The Swedish coastal zone monitoring programme. In: Kamphorst, E., Jørgensen, L.F., van Griensven, A. and Vanrolleghem, P.A. (Eds.) *Joint Use of Modelling and Monitoring for Implementing the Water Framework Directive.* Workshop report, 1st workshop: State of the art on existing monitoring programmes around Europe. Harmoni-CA. http://www.harmoni-ca.info/

Jørgensen, L.F., Brandt, G. and Vandenberghe, V. (Eds.) (2006). *Joint Use of Modelling and Monitoring for Implementing the Water Framework Directive.* Workshop report, 3rd workshop: Case studies. Harmoni-CA. http://www.harmoni-ca.info/

Jørgensen, L.F. (2005). Groundwater monitoring in Denmark. In: Kamphorst, E., Jørgensen, L.F., van Griensven, A. and Vanrolleghem, P.A. (Eds.) *Joint Use of Modelling and Monitoring for Implementing the Water Framework Directive.* Workshop report, 1st workshop: State of the art on existing monitoring programmes around Europe. Harmoni-CA. http://www.harmoni-ca.info/

Kamphorst, E., Jørgensen, L.F., van Griensven, A. and Vanrolleghem, P.A. (Eds.) (2005). *Joint Use of Modelling and Monitoring for Implementing the Water Framework Directive.* Workshop report, 1st workshop: State of the art on existing monitoring programmes around Europe. Harmoni-CA. http://www.harmoni-ca.info/

Klauer, B. and Brown, J.D. (2003). Conceptualising imperfect knowledge in public decision making: ignorance, uncertainty, error and 'risk situations'. *Environmental Research, Engineering and Management* **27**, 124–128.

Koussis, A.D. (2005). On the status of monitoring the water resources of Greece. In: Kamphorst, E., Jørgensen, L.F., van Griensven, A. and Vanrolleghem, P.A. (Eds.) *Joint Use of Modelling and Monitoring for Implementing the Water Framework Directive.* Workshop Report, 1st workshop: State of the art on existing monitoring programmes around Europe. Harmoni-CA. http://www.harmoni-ca.info

Larsen, S.E., Jensen, C. and Carstensen, J. (2002). *Statistisk optimering af monitering-sprogrammer på miljøområdet.* Eksempler fra NOVA 2003. Danmarks Miljøunder-søgelser. 196 pp. Faglig rapprt fra DMU nr. 426. In Danish with English Summary.

Pahl-Wostl, C. (2002). Towards sustainability in the water sector – the importance of human actors and processes of social learning. *Aquatic Science* **64**, 394–411.

Pataki, B. (2005). Grondwater monitoring in Hungary. In: Kamphorst, E., Jørgensen, L.F., van Griensven, A. and Vanrolleghem, P.A. (Eds.) *Joint Use of Modelling and Monitoring for Implementing the Water Framework Directive.* Workshop report, 1st workshop: State of the art on existing monitoring programmes around Europe. Harmoni-CA. http://www.harmoni-ca.info/

Purvina, M. (2005). Surface water monitoring programme in Latvia. In: Kamphorst, E., Jørgensen, L.F., van Griensven, A. and Vanrolleghem, P.A. (Eds.) *Joint Use of Modelling and Monitoring for Implementing the Water Framework Directive.*Workshop report, 1st workshop: State of the art on existing monitoring programmes around Europe. Harmoni-CA. http://www.harmoni-ca.info/

Refsgaard, J.C. (1996). Terminology, modelling protocol and classification of hydrological model codes. In: Abbott, M.B. and Refsgaard, J.C. (Eds.) *Distributed Hydrological Modelling.* Kluwer Academic Publishers, Dordrecht, The Netherlands, pp. 17–39.

Refsgaard, J.C., Thorsen, M., Jensen, J.B., Kleeschulte, S., and Hansen, S. (1999). Large scale modelling of groundwater contamination from nitrogen leaching. *Journal of Hydrology* **221**(3–4), 117–140.

Refsgaard, J.C. and Henriksen, H.J. (2004). Modelling guidelines – terminology and guiding principles. *Advances in Water Resources* **27**(1), 71–82.

Refsgaard, J.C., Henriksen, H.J., Harrar, W.G., Scholten, H. and Kassahun, A. (2005a). Quality assurance in model based water management – Review of existing practice and outline of new approaches. *Environmental Modelling & Software* **20**, 1201–1215.

Refsgaard, J.C., Nilsson, B., Brown, J., Klauer, B., Moore, R., Bech, T., Vurro, M., Blind, M., Castilla, G., Tsanis, I. and Biza, P. (2005b). Harmonised techniques and representative river basin data for assessment and use of uncertainty information in integrated water management (HarmoniRiB). *Environmental Science and Policy* **8**, 267–277.

Refsgaard, J.C., Nilsson, B. and Bache, K. (2006) *HarmoniRiB Final Report.* Geological Survey of Denmark and Greenland, Copenhagen. http://workplace.wur.nl/harmonirib.

Refsgaard, J.C., van der Sluijs, Højberg, A.L. and Vanrolleghem, P.A. (2007) Uncertainty in the environmental modelling process – A framework and guidance. *Environmental Modelling & Software* **22**, 1543–1556.

Scholten, H., Kassahun, A., Refsgaard, J.C., Kargas, T., Gavardinas, C, and Beulens, A.J.M. (2007). A methodology to support multidisciplinary model-based water management. *Environmental Modelling & Software* **22**, 743–759. doi:10.1016/j.envsoft.2005.12.025

Steenstra, M., Troch, P., Santbergen, L. and Jørgensen, L.F. (Eds.) (2004). *Remote sensing and data assimilation techniques*. Workshop report. Harmoni-CA, Deliverable no. D4-2. http://www.harmoni-ca.info/

Thorsen, M., Refsgaard, J.C., Hansen, S., Pebesma, E., Jensen, J.B. and Kleeschulte, S. (2001). Assessment of uncertainty in simulation of nitrate leaching to aquifers at catchment scale. *Journal of Hydrology* **242**, 210–227.

UNECE (1998). *The United Nation Economic Commission for Europe. Convention on Access to Information, Public Participation in Decision-Making and Access to Justice in Environmental Matters*. Aarhus Denmark. http://www.unece.org/env/pp/documents/cep43e.pdf

United Nations (2000) *The Aarhus Convention: An Implementation Guide*. New York and Geneva. The Regional Environmental Center for Central and Eastern Europe & the Danish Environmental Protection Agency. Available at: http://www.unece.org/env/pp/acig.pdf

Van Loon, E. and Refsgaard, J.C. (Eds.) (2005). *Guidelines for assessing data uncertainty in hydrological studies*. HarmoniRiB Report. Geological Survey of Denmark and Greenland. http://www.harmonirib.com

WMO (1994). *Guide to Hydrological Practices*. Fifth Edition, WMO – No. 168. World Meteorological Organisation, Geneva, Switzerland.

Zupan, M. (2005). Surface water and groundwater monitoring in Slovenia. In: Kamphorst, E., Jørgensen, L.F., van Griensven, A. and Vanrolleghem, P.A. (Eds.) *Joint Use of Modelling and Monitoring for Implementing the Water Framework Directive*. Workshop report, 1st workshop: State of the art on existing monitoring programmes around Europe. Harmoni-CA. http://www.harmoni-ca.info/

8. APPENDIX – HARMONIRIB RIVER BASINS

Odense, Denmark

Summary of basin and key water management issues

The Odense Fjord catchment is situated at the island Fyn in the middle of Denmark. The catchment area is 1,046 km^2. Odense is one of the Pilot River Basins under the WFD Common Implementation Strategy. The landscape is moraine formed by the last glaciation. The topography is characterised by low rolling hills. All water supply is based on groundwater from mainly confined sandy aquifer overlain by till.

The main problem with respect to surface water quality is the eutrophication in the Odense Fjord caused by nutrients originating from non-point sources. In addition, the groundwater quality is many places threatened by point and non-point sources. These problems have generated significant conflicts of interests between the agricultural sector and other stakeholders.

Further descriptions of the catchment, the water management issues and the data can be found in HarmoniRiB reports (http://workplace.wur.nl/harmonirib) and on the Odense PRB website (http://odenseprb.ode.mim.dk/english/).

Data availability

As indicated from Table 3.2 the data availability is generally very good for Odense. In particular there are many data on quality aspects, both for surface water and groundwater.

Data accessibility

Data are generally easily accessible from national and water authority databases in digital formats. The water authorities (Fyns Amt until 2006, Ministry of Environment and Odense Municipality from 2007) have access to all the available data. Data access for third parties, e.g., for research purposes is possible. The main potential obstacles in this respect are that meteorological data and topographical data not are available for free. For research purposes it is possible to make ad hoc arrangement with the two organisations controlling these data.

A substantial part of the data has been made available for the HarmoniRiB database, from where it can be made freely available for non-commercial research purposes.

Modelling studies and research projects

The water authority and some of the stakeholders, in particular the local water supply company Odense Water Ltd, have during the past 10–15 years established a large number of comprehensive models. The modelling work has usually been outsourced to consulting firms. In addition, the Odense river basin has been used as a site in many national and European research projects such as Baseline, Euroharp, HarmoniRiB and Bridge with research institutes as the main executing partners as well as LIFE and INTERREG projects executed by the water authority itself.

Most of the modelling studies have been carried out by use of the existing data. However, for some of the research studies aiming at providing new insights into process descriptions dedicated data collection campaigns were carried out. As can be seen from Table 3.3 the modelling studies cover water availability, surface water and groundwater quality with rather complex models.

Geropotameau, Greece

Summary of basin and key water management issues

The Geropotamou basin covers an area of 600 km^2 and is located in the central southern part of Crete, about 50 km south of the city of Heraklion. The basin faces a severe groundwater resources problem due to overexploitation of the aquifer, mainly for irrigation of 250 km^2 of agricultural land located in the Messara Valley. This situation has led to Messara Valley being threatened by desertification. A dramatic drop of about 20 m in the mean annual groundwater level during the last 30 years has reduced the available water due to uncontrolled use and created tension amongst the users. The groundwater level dropdown started with the introduction of pumping of groundwater for drip-irrigation of the main crop which is olive trees. Following that, pumping became uncontrolled as farmers established an unknown number of unregistered wells. Illegal use of groundwater water is difficult to police.

Further descriptions of the catchment, the water management issues and the data can be found in HarmoniRiB reports (http://workplace.wur.nl/harmonirib).

Data availability

As indicated in Table 3.2 Geropotameau has many data for meteorological, river discharge and groundwater level, but very limited data on surface water and groundwater quality.

Data accessibility

Some data are easily accessible from national and water authority databases in digital formats. The water authority (Water Resources Department) has access to all the available data. Data access for third parties, e.g., for research purposes is possible, but need to be negotiated on an ad hoc basis.

A substantial part of the data has been made available for the HarmoniRiB database, from where it can be made freely available for non-commercial research purposes.

Modelling studies and research projects

No modelling studies have been carried out to support the water resources management in practise. During the HarmoniRiB project a resource assessment modelling study was carried out comprising rainfall-runoff (Sacramento) and groundwater (MODFLOW) conditions.

Jucar, Spain

Summary of basin and key water management issues

The Júcar River Basin comprises all river basins which run into the Mediterranean sea, between the left bank of "La Gola del Segura", in its mouth, and the mouth of the Cenia river, as well as the endorreic basin of Pozohondo. The total area is approximately 43,000 km^2. The Jucar RBD is mainly influenced by the Mediterranean climate, which is characterised by an intense and long dry summer period. The Jucar River Basin can be defined as a semi-arid or even arid zone as a whole. Jucar is one of the Pilot River Basins under the WFD Common Implementation Strategy.

The main water use in the basin is irrigation for agriculture. The main source is the groundwater, because surface runoff is not sufficient to satisfy the water demand. Other uses are urban use, hydropower, tourism and ecological needs for preserving river discharge and wetlands. The main water resources problem is the unsustainable use of groundwater for irrigation, which during the past decades many places has resulted in lowering of groundwater levels of tens of meters.

Further descriptions of the catchment, the water management issues and the data can be found in HarmoniRiB reports (http://workplace.wur.nl/harmonirib).

Data availability

As indicated in Table 3.2 Jucar has many data for meteorological, river discharge and groundwater data, and some data on surface water and groundwater quality.

Data accessibility

Some data are easily accessible from national and water authority databases in digital formats. The water authority have access to all available data. Data access for third parties, e.g., for research purposes is possible, but need to be negotiated on an ad hoc basis. In this respect meteorological data need particular attention.

A substantial part of the data has been made available for the HarmoniRiB database, from where it can be made freely available for non-commercial research purposes.

Modelling studies and research projects

Previously the following types of modelling studies have been carried out: rainfall-runoff modelling, hydro-ecological modelling and groundwater modelling (MODFLOW).

Candelaro, Italy

Summary of basin and key water management issues

The Candelaro river catchment is located in the northern part of the Apulian region in the geo-morphologically charateristic environment of the Tavoliere of Apulia. It covers 2,330 km^2.

The climate is Mediterranean and the rainfall rates have decreased about 30% during the last 40 years. During the same period the groundwater exploitation, mainly for irrigation of grapes and vegetables, has increased dramatically.

As a result of this parts of the Candelaro basin are threatened by desertification. Some place, a drop in groundwater levels up to 20 m has occurred during the last 20 years of the last century. Management of groundwater resources is difficult due to uncontrolled use by farmers illegally using unregistered wells. An allegedly oversized dam has been constructed and is currently filled up, aiming to fulfil the irrigation water needs in the basin.

There are also concerns on the water quality.

Further descriptions of the catchment, the water management issues and the data can be found in HarmoniRiB reports (http://workplace.wur.nl/harmonirib).

Data availability

As indicated in Table 3.2 Candelaro has many data for meteorological, river discharge and groundwater level, but very limited data on surface water and groundwater quality.

Data accessibility

Some data are easily accessible from national and water authority databases in digital formats. The regional water and environmental authority has access to all the available data. Data access for third parties, e.g., for research purposes is generally possible on an ad hoc basis. In this respect meteorological data need particular attention.

A substantial part of the data has been made available for the HarmoniRiB database, from where it can be made freely available for non-commercial research purposes.

Modelling studies and research projects

A large number of modelling studies have previously been carried out both to support the water resources management and for research purposes. The types of modelling include hydrological water balance studies, leaching of nutrient (SWAT) and groundwater modelling.

Vecht, The Netherlands

Summary of basin and key water management issues

The Vecht river basin covers an area of 3,780 km^2 of which 1,980 km^2 is located in The Netherlands and the remaining in Germany, where the river originates. The Vecht catchment is part of the Rhine catchment.

Land use in the southern part is predominantly intensive animal husbandry, with growing of grass and maize. In the northern part there is more arable land, growing mainly potatoes. The main purposes of water use are: agriculture, drinking water, and recreation. The human pressure on the aquatic environment is high, both from cities and from intensive agriculture. Discharges from many of the sewage treatment plants are into relatively small waters. Most of the waters in the catchment have been strongly regulated by normalisation and dams. In large parts of the area water inlet from outside the catchment plays an important role for agriculture in the summer.

Further descriptions of the catchment, the water management issues and the data can be found in HarmoniRiB reports (http://workplace.wur.nl/harmonirib).

Data availability

As indicated in Table 3.2 the data availability is generally very good for Vecht. In particular there are many data on quality aspects, both for surface water and groundwater.

Data accessibility

Data are generally easily accessible from national and water authority databases in digital formats. The water authorities have access to all available data. Data access for third parties, e.g., for research purposes is possible, but ad hoc agreements are some times required.

A substantial part of the data has been made available for the HarmoniRiB database, from where it can be made freely available for non-commercial research purposes.

Modelling studies and research projects

The water authority has during the past years established several models to support water management. In addition, the Vecht river basin has been used as a site in many national and European research projects such as Euroharp and HarmoniRiB with research institutes as the main executing partners.

Thames, England

Summary of basin and key management issues

The River Thames has a catchment area of around 13,000 km. This includes the main Thames and tributaries such as the Kennet and Cherwell. It covers much of the rural areas of the counties of Wiltshire, Oxfordshire and West Berkshire, and urban areas such as Oxford, Newbury, Reading, Slough, Luton and London. There are no major natural lakes within the Thames Basin, but there are large reservoirs for water supply. The geology is varied. Chalk is the most common hydrogeological formation contributing to water supply. Other aquifers are the Lower Greensand, Blackheath Beds, Bagshot Beds, the Jurassic limestones of the Cotswold hills, and sandy and gravelly drift deposits.

The catchment has a growing population of 12 million, and is under severe pressure to provide further living space, water resources and waste disposal. A major issue centres around managing abstraction to meet demands, without detriment to the ecological functioning of riverine habitats. Many sewage treatment works discharge into the river, with associated water quality issues.

Further descriptions of the catchment, its water management issues and data can be found in the HarmoniRIB reports (http://workplace.wur.nl/harmonirib).

Data availability

As indicated in Table 3.2, water recourses management, surface water quality and ground water quality data availability for the Thames Basin is extensive. However, there are not much agriculture data available for the basin.

Data accessibility

In general, data are easily accessible from research councils and government water authorities. Most of the data are available digitally. The primary source for river flow data is the National River Flow Archive held at the Centre for Ecology and Hydrology in Wallingford. Groundwater level and chemistry can be obtained from the National River Flow Archive or the British Geological Survey. River water chemistry and ecological data may be obtained from the Environment Agency. Weather data are available from the Met Office.

A substantial part of the data is available for the HarmoniRIB database, from where it is freely available for non-commercial research purposes.

Modelling studies and research projects

A wide range of modules were developed for the study of the Thames Basin. Modelling is primarily used for the studies of flood forecasting and water quality. Today most of the modelling is overseen by the Environment Agency.

Svratka, Czech Republic

Summary of basin and key water management issues

The Svratka river basin is situated in the Moravian part of the Czech Republic and is part of the Danube river basin. The catchment area is 4,115 km^2. The basin is part of the Ceskomoravska vrchovina highland and the main geomorphology are deep and fault igneous rocks, the rest are Perm and Carboniferous structures, unconsolidated Tertiary structures, flysch belts and Quaternary structures.

The key water management issues in the basin are flood protection and quality of surface water bodies. Fast development of the settlement areas in the floodplain during the last century has created problems with flooding in many towns and subsequently requests for long time sustainable flood protection in the basin. Another important issue is the water eutrophication and blue green algae development mainly in the water reservoir Brno on Svratka river resulting in heavy impact on the recreation and water use.

Further descriptions of the catchment, the water management issues and the data can be found in HarmoniRiB reports (http://workplace.wur.nl/harmonirib).

Data availability

As indicated in Table 3.2 the data availability is good for the Svratka river basin. The problem is mainly the lack of biological data.

Data accessibility

The data are generally accessible from state institutions (like Czech Hydrometeorological Institute or Water Research Institute) or from Water Boards. Data are free of charge for governmental, regional and local authorities; other users have to pay for data.

A substantial part of the data is available for the HarmoniRiB database, from where it can be made freely available for non-commercial research purposes.

Modelling studies and research projects

Several modelling studies focussing on water management problems in the basin, mainly on flood protection (regional studies of possible flood protection measures) and also regarding quality of surface water (project Clean Svratka river basin) have been carried out in the area.

Weisse Elster, Germany

Summary of basin and key management issues

The Weisse Elster river basin is a subcatchment of the Saale, which is the second largest tributary of the Elbe river. The catchment area is about 5,300 km^2 and is situated in the German federal states of Saxony, Thuringia and Saxony-Anhalt, with its source being in the Erzgebirge (Ore Mountains) in the Czech Republic. The upper part of the basin is mountainous with igneous and metamorphic rocks as well as consolidated tertiary rocks (sandstone) being the main geological features. It is characterized by steep slopes and narrow valleys with hardly any floodplains. Groundwater resources are scarce. The lower part of the basin is situated in the lowlands and mainly consists of Pleistocene coverage. The valleys are broader with extensive floodplains and there are substantial groundwater resources. Precipitation varies between 500 mm in the northern part of the basin (lowlands) and 1000 mm in the southern part (mountains) with annual runoff varying between 50 and 800 mm.

Due to several reservoirs and numerous hydraulic structures, the hydrological situation in the catchment is very complex. Water quantity and quality management are closely related to the various economic activities in the river basin. Due to extensive construction, diffuse pollution from agriculture, industrial and municipal discharge and open pit mining waters, the Weisse Elster is considered biologically and chemically heavily polluted. Although water quality has substantially improved in the last few years, a comprehensive remediation programme is necessary to meet the WFD targets.

Further descriptions of the catchment, its water management issues and data can be found in the HarmoniRIB reports (http://workplace.wur.nl/harmonirib).

Data availability

As indicated in Table 3.2 the data availability is generally very good for the Weisse Elster. In particular there are many data on water resources assessment and surface water aspects.

Data accessibility

In general, data are easily accessible from research councils and government water authorities. Most of the data are available digitally. The water authorities have access to all available data. Data access for third parties, e.g., for research purposes is possible, but ad hoc agreements are some times required.

A substantial part of the data has been made available for the HarmoniRIB database, from where it is freely available for non-commercial research purposes.

Modelling studies and research projects

A wide range of modelling studies has been carried out both for water management and for research purposes. The modelling studies have covered most water resources issues with a focus on surface water aspects.

Guidance Report II.3

Good practice in joint use of monitoring and modelling

Anker Lajer Højberg, Jens Christian Refsgaard, Lisbeth Flindt Jørgensen, Frans van Geer and István Zsuffa

Preface

In order to support the implementation of the Water Framework Directive, the European Commission established a cluster on Integrated Catchment Modelling (CatchMod). The objective of this cluster was the development of common harmonised modelling tools and methodologies for integrated water management at river basin or sub-basin scales, including the interface to the coastal zone.

Within the CatchMod cluster, the project Harmoni-CA – Harmonised Modelling Tools for Integrated River Basin Management – had the objective to create a forum for unambiguous communication, information exchange and harmonisation of the use and development of ICT-tools relevant to integrated river basin management, and the implementation of the WFD. Harmoni-CA was a large-scale concerted action, meaning that it does not carry out research, but provides synthesised available knowledge with the help of knowledge providers such as researchers. Typical actions of Harmoni-CA were meetings and workshops, leading to synthesis reports and guidances.

One of the tasks of the Harmoni-CA project is defined within Harmoni-CA Work Package 4: "Joint use of monitoring and modelling". The objective of this WP was to develop a guideline on best practise on joint use of the two disciplines, providing an overview of possibilities and a number of examples. To accomplish this task five workshops have been organised in the period 2004 to 2007 under the umbrella "Joint use of monitoring and modelling when implementing the Water Framework Directive". The important goals of the workshops have been to facilitate discussions and exchange experiences between the researchers/developers and practitioners to bring the two communities closer and to develop a common understanding on the needs and opportunities. The workshops succeeded to bring almost 80 people together representing 21 of the 27 EC countries besides participants from Norway, Moldova, Ukraine and the US. The weighting between researchers and water managers was equal, a few representing both sides. Besides, a couple of consultants attended the workshops, mostly presenting tangible examples from all around the world. Several attended more than one of the workshops. Experiences, conclusions, recommendations and lessons learned are brought forward in this Guidance Report.

 Integrated Assessment for Water Framework Directive Implementation Water Framework Directive Series Volume 2. Edited by Peter Vanrolleghem. ISBN: 9781843393269. Published by IWA Publishing, London, UK.

An earlier version of this document was presented and discussed at the Harmoni-CA review workshop in Cologne 1 March 2007. The reviewers were Faycal Bourai, Morten Sørensen and Bernardas Paukstyks.

Keywords
Data Assimilation, Data Quality Assurance, Experimental Design, Mathematical Modelling, Monitoring Network, River Basin Management

1. EXECUTIVE SUMMARY

What do monitoring data tell us about the environment? How do we formulate programmes of measures based on monitoring data? How confident are we in the decisions we make based on observation data?

In the Water Framework Directive much emphasis is put on monitoring data and the guiding principle is based on uncertainty and risk assessments. Monitoring is a prerequisite for water management but data itself only reflect the current state of the system at the point and time where the measurements were taken. A challenging task is thus to move from the discrete measurements to integrated water resources assessment, where the risk of making the wrong decisions are at an acceptable level.

In accordance to this, the need to support the EU environmental monitoring by new knowledge and scientifically based tools including an enhanced communication, is often emphasised by specialists and even by practitioners. However, descriptions of the present state-of-the-art in joint use of modelling and monitoring techniques show that most often models are not considered an option when the monitoring obligation in the Water Framework Directive are solved in practise. If modelling is used the interaction is commonly reduced to a single step one-way (data-to-model) communication.

Practical experience from several modelling studies and research projects have demonstrated that joint use of monitoring and modelling has a much larger potential than what is presently utilised. One of the main arguments for using modelling tools is to bring the monitoring data "alive". Although monitoring data may be subject to rigid quality assurance protocols during sampling and analysis, many errors in data may only be discovered if they are interpreted based on a conceptual understanding of the system and confronted with other data types, describing the same system only from another "perspective". To formulate future strategies, such as programmes of measures or redesign of monitoring networks, a thorough understanding of the natural system is a prerequisite. This can only be achieved if we are able to understand all data from the system based on a coherent and plausible conceptual understanding. In this sense, model systems provide a powerful tool to integrate all data and to test and revise our understanding. Additionally, models have the ability to test different scenarios. By this, the effect of different alternative programmes of measures can be tested prior to implementation and may thus aid in the identification of the most effective strategy.

In the present report, some methods and examples are provided on how modelling and monitoring can be used jointly, with emphasis on the tasks required in the WFD monitoring programmes:

- *Quality assurance.* Models are powerful tool to integrate all data from a system and analyse their interconnection. Inconsistency and errors in data may only be recognised by such analyses.
- *Inter- and extrapolation.* Point measures must be converted into temporal and spatial continuous data. Using a model based on a conceptual understanding, inter- and extrapolation becomes knowledge based as opposed to the generic inter- and extrapolation routines.
- *Conceptual understanding.* A conceptual understanding is a prerequisite for all management plans. Models can be used to formulate and test such conceptual understanding.
- *Anthropogenic activities.* Effects of anthropogenic activities may be difficult to recognise from monitoring data alone. Models may be used to filter out natural variations and thereby amplify the signals from the anthropogenic activities.
- *Design of monitoring network.* Monitoring network can be optimised by the use of models to evaluate the information value of data types and monitoring location points for an existing or new monitoring programme.

The Harmoni-CA workshops on joint use of monitoring and modelling identified several obstacles limiting the present joint use, of which many can be related to historical praxis. However, the workshops also demonstrated that if people from the monitoring community and the modelling community are brought together for open-minded discussions and with free hands they are very open to discuss ideas and suggestions from the other side of the table, and able to see the perspectives in the combined use. To advance the joint use of monitoring and modelling there appears to be a strong need to first of all raise the awareness of how monitoring and modelling can be combined. Additionally, a shift in the perception of models is needed, where models currently often are considered useful only to provide answers to specific questions. This view must be replaced by recognition of models as being valuable tools that should be used continuously and interactively along with monitoring data, a tool that serves many purposes from the design phase of monitoring programmes to the interpretation and quality assurance of the monitoring data.

The WFD imposes many new challenges and among those are the requirements to consider an integrated use of all waters and an evaluation of the qualitative status in terms of ecological aspects. To fulfil these requirements development of new tools and models is a necessity. To avoid a future parallel development of monitoring and modelling praxis in the field of integrated and ecological assessments, it is of vital importance that monitoring and modelling are realised to be inter-linked activities and not independent disciplines.

Within this work two workshops have been carried out, where six real life case studies from different areas around Europe have been elaborated to create flow charts on how to implement the WFD, considering when to monitor and when to model and where these two disciplines should join forces. These flow charts have been combined to a common flow chart that we believe to be applicable in most WFD implementation challenges around Europe. The flow chart is described in the Appendix.

2. INTRODUCTION

In the past, monitoring has traditionally been considered an independent discipline, but within the last decades modelling has entered the arena as a supplementary tool to help extracting the information retained in the observation data. In the research communities it is generally accepted that monitoring and modelling are inter-linked activities (Holt *et al.*, 2000; Parr *et al.*, 2003). The guiding principle for monitoring in the WFD is based on uncertainty and risk assessments, which is quite advanced, when considered as guidance to practitioners in water resources management. In full accordance with this line of thinking the guidance document on planning (EC, 2003b) describes the benefits of using models to support the WFD implementation. The WFD therefore creates new challenges on monitoring and modelling and provides new incentives for improving the joint use of monitoring and modelling (Højberg *et al.*, 2007; Zsuffa *et al.*, 2005). The need to support the EU environment monitoring, specific to the groundwater domain, by new knowledge and scientifically based tools including an enhanced communication among scientists, stakeholders and policy-makers is discussed by Quevauviller (2005). Descriptions of present state-of-the-art in joint use of modelling and monitoring techniques, however, show that most often models are not considered an option when the monitoring obligation in the WFD are solved in practise (Kamphorst *et al.*, 2005; Arustiene *et al.*, 2005; Borowski and Hare, 2007).

2.1. Why joint use?

The prospectives of models have varied since the early days of the modelling discipline ranging from deep scepticism to overly optimism. Critics often argue that a model is only a speculative picture of the natural system, and as such does only provide speculative information of the system of interest. This argument may hold true if the model's ability to mimic the system under concern has not been tested. Models may be thought of as sophisticated databases that provide an ordered way to store field data and define the relationships between data, e.g., the temporal development in streamflow as a response to a storm event. These relationships may be based on pure empiricism, statistical relations, or mechanistic processes, but are all generic formulations and models must therefore be adjusted to describe the site-specific conditions. The adjustment of model parameters is commonly accomplished through calibration, or history matching, where model parameter values are varied until the model reproduces the response of the reality within some predefined accuracy. If no site-specific data are available, the models can not provide insight in the system of interest, but are limited to provide some general insight in the system behaviour, e.g., how different processes interact. Using such models to predict the state of a site-specific natural system and its response to stresses will be guesswork and monitoring data are therefore a prerequisite to site-specific modelling.

Data alone do, on the other hand, only provide information of the current state of the environment, e.g., a chemical concentration, at a specific location and at the time at which the data was collected. This information is, in itself, of very little use. If more measures of the compound have been taken a time series can be constructed and obvious trends in the concentration levels may be revealed. The data do, however, not explain the cause for this trend. To extract this kind of information from the data set one needs to interpret the data based on an understanding of the physical system, that is, we formulate a conceptual understanding (or conceptual model) of the system from which we explain the data. As such we add knowledge to the data, which is necessary if we want to formulate future management strategies based on the observations. Natural systems are, however, very complex with several processes occurring simultaneously and interacting. It is therefore very difficult to fully comprehend and separate the different processes, their interaction and their consequences by interpretation of field data alone. In this context, models can be seen as a formal description of knowledge of the hydrological system studied, describing the spatial and temporal coherence in the system.

Presently, the main and often only interaction between monitoring and modelling is in the phases of constructing, calibrating and validating models.

Once these tasks have been completed, the monitoring data are, at best, only used to quantify the uncertainty of the models, by some expression of the deviation between the model prediction and observed variables. The interaction is thus merely a single step one-way (data-to-model) link. Much improvement in models may, however, be obtained if the quantified uncertainty is used as basis for correcting and up-dating the model and model predictions. An example of such procedure is data assimilation, which is nearly only applied in flood forecasting but may be used in other domains as well for model updating.

Monitoring data and models provide complementary information of the system. The model captures physical knowledge about the systems behaviour and monitoring provides information from the actual system including phenomena which are not included in the model. It may be argued that "modelling without monitoring data is guesswork" and "monitoring without models is a waste of money".

2.2. What are the problems?

Several obstacles have been brought forward for the joint use of monitoring and modelling in the practical world. From the workshops the following impediments were identified:

- *Lack of awareness of what models can do.* For many implementers of the WFD it has been unclear what is actually required to fulfil the directive. This uncertainty has not encouraged the exploitation of new methods or strategies, and as the environmental state has traditionally been examined by means of monitoring data in many countries, this is still the primary choice.
- *Lack of confidence in models.* Model simulation is inherently uncertain as the natural system can never be described in a complete and exact way. By applying models it becomes possible to assess and display the uncertainty. While this is considered a true benefit in the research world it causes some decision makers to be reluctant in the use of models. This was the focus of another workshop in Harmoni-CA (Brugnach *et al.*, 2007).
- *Technical background.* Modellers and monitoring people often belong to different institutes and have different technical background. In most cases they do not work together on a daily basis and are not aware of the mutual benefits they may gain by close collaboration.
- *State of mind.* People from the monitoring community tend to state that a model reflects the (subjective) mind of the modeller and stick to their real data – nothing is as firm and trustworthy as highly valued and long time

series. On the other hand modellers state that data are too noisy and are not worth anything, if we do not understand the mechanisms – how data are linked physically, mechanically and chemically – the data rich-information poor syndrome.

- *Conflict of interests.* There might be a clash of interests when it comes to the available economical resources. People from the monitoring community are under a constant pressure on their resources in terms of questions from decisions makers as whether the activities in data collection are carried out with enough efficiency; are too much data collected, should the programmes be put on the market etc. If the monitoring community senses that the modelling activities must be financed from their monitoring budgets they are already opposed to the idea, not from a technical point of view but simply in terms of survival.
- *Lack of skill.* For some member states the modelling discipline is juvenile and few have been trained as modellers. When new modellers have been educated they are often hired in private consultancies and do not work closely with the water mangers.
- *Lack of time and resources.* Water managers are often faced with problems that must be solved within a short timeframe and in a cost-effective manner. Models are often time-consuming and expensive to develop and many water managers are hesitant to employ models as they are dubious if it is worthwhile.
- *Lack of data.* A commonly posed argument is that there is never enough data for modelling purposes. This includes both inadequate spatial and temporal resolutions but also restrictions in data access, where useful data may exists but are either difficult or very expensive to acquire. A common problem is that modellers are not included in the design of monitoring programs which may then be designed with a too narrow scope of monitored parameters for modelling purposes. The data availability and accessibility issue is further discussed in Refsgaard *et al.* (2010).
- *Incompatibility of time scales.* Monitoring often takes many years before meaningful information is obtained, whereas most models are built to solve a problem at present. While monitoring should be designed to focus on future problems, model construction, calibration and validation needs data from the past.

Common to all the obstacles are that they are not related to technical difficulties. The scientific basis to improve the combined use is well developed and new methods are continuously developed.

2.3. Objectives and target audience of this document

The obstacles listed in the previous section need to be solved at different levels and by different means. Most of the obstacles may be tackled at an institutional level, and different solutions strategies may be necessary to solve the problem at different institutions. But before a solution is sought the motivation for using models must exist – is it worthwhile? The first obstacle to tackle is thus to raise awareness of the mutual benefits. This was the primary objective of the workshops and here it was experienced that if people from the monitoring community and from the modelling community are brought together for open-minded discussions and with free hands they are very open to discuss ideas and suggestions from the other side of the table. Monitoring people ask for tangible examples showing benefits and limitations and modellers asks for information on data and for being involved in possible redesigning monitoring programmes.

The objective of the present report is thus to demonstrate some of the possibilities on how to combine modelling and monitoring and to illustrate the advantages thereof. Many advantages of the combined use are natural spin offs from a model study, which do not require additional work beyond the modelling process. For more comprehensive interaction it may be necessary to employ different methods, which vary widely from very simple tasks that can be carried out without much additional effort, to more complex methods where the workload is substantial. However, the present development and practical experiences offers a wide range of opportunities to combining monitoring data and modelling in a more intelligent way than is commonly praxis.

The combined use of monitoring and modelling may be beneficial in wide range of studies independent of the study objectives. In the present report only aspects related to the implementation of the WFD are discussed, however, the discussions and methods presented are not limited to the implementation of the WFD but are generic, and may thus apply to any study.

The report is not a technical guideline on specific methods that can be employed in the effort of combining monitoring and modelling. Instead, the focus is to illustrate the benefits of the combined use, through a short description on the basic principles and benefits of different topics, where the joint use already are or easily can be included, as well as some methods that are still in the developing phase, but in the near future may be operational.

The target audience is policy makers, water managers and professionals involved in implementation of the Water Framework Directive.

3. MONITORING AND MODELLING WITHIN THE WFD

The starting point for discussing the potentials of joint use of monitoring and modelling in the implementation of the WFD is an identification of the different monitoring programmes and their specific objectives. Based on the specific objectives and requirements for a given study different modelling approaches may be appropriate.

3.1. WFD monitoring programmes

The overall objective of the Water Framework Directive is to achieve a good quantitative and qualitative status of all waters by 2015. To fulfil the goals of the WFD the EU member states are obliged to prevent further deterioration, protect and enhance the status of the water resources. The WFD thus prescribes both an improvement for waters already negatively affected by anthropogenic activities, as well as their protection with respect to future activities.

To accomplish the overall aim of the WFD a number of tasks have to be carried out. Firstly, all the water bodies must be described, delineated and classified with respect to the quantitative and qualitative status. This classification includes the detection of possible trends that may cause the water body to be at risk of failing to achieve the good status in the future. Next, the cause of the trend has to be assessed by separating the anthropogenic activities from natural variation, and the deviation of observed conditions to those that would be found under reference conditions must be evaluated. Finally, pressures have to be identified and evaluated upon the possible risk they may pose to the water bodies. With an overview of the present status and the possible future development, programmes of measures have to be established and the effect thereof evaluated. The backbone to resolve these issues within the WFD is the establishment of comprehensive monitoring networks. Four types of monitoring has been defined, with different purposes as described in the monitoring guidance (EC, 2003a):

- *Surveillance Monitoring*: To determine status; to supplement and validate the initial impact assessments; to assess long time changes; to validate hypotheses and to assist in designing future monitoring programmes.
- *Operational Monitoring*: To assess status of water bodies at risk of failing to meet environmental objectives; to assess effects of programmes of measures. Operational monitoring must be carried out in all bodies receiving priority substances.

- *Investigative Monitoring (only surface water)*: Where the reason for any exceedance is unknown; where good status is not likely to be achieved and where operational monitoring has not already been established; to ascertain the magnitude and impacts of accidental pollution; Measure only the problem-defined variable(s).
- *Water Level Monitoring Network Design (only groundwater)*: For groundwater quantitative monitoring is required. This should be designed to ensure that the available groundwater resource is not exceeded by the long-term annual average rate of abstraction; to ensure that abstractions and other anthropogenic alterations to groundwater levels do not adversely affect associated surface water bodies and terrestrial ecosystems that depend directly on groundwater with respect to their water needs; to ensure that anthropogenic alterations to flow direction must not have caused, or be likely to cause, saltwater or other intrusion.

The relation between the different monitoring programmes and the requirements in the WFD (Annex II analysis, Article 5 and 8) is shown schematically for surface water and groundwater monitoring in Figures 3.1 and 3.2, respectively.

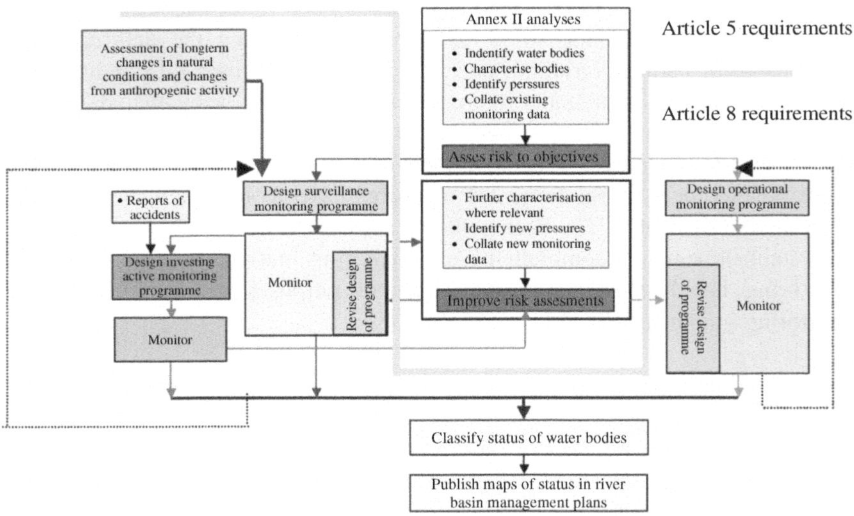

Figure 3.1 Schematic diagram illustrating the relationship between the WFD requirements and the surface water monitoring programmes (from EC, 2003a)

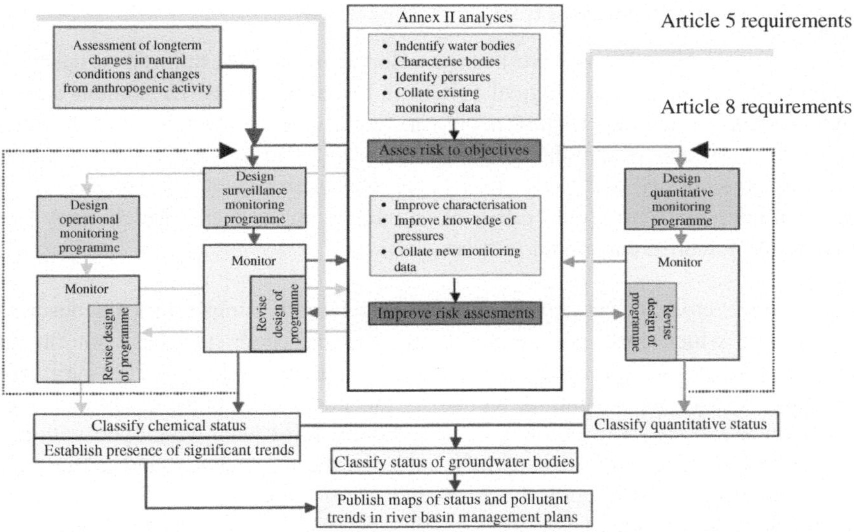

Figure 3.2 Schematic diagram illustrating the relationship between the WFD requirements and the groundwater monitoring programmes (from EC, 2003a)

The overall purpose of the monitoring programme is to establish a coherent and comprehensive overview of the water status within each river basin district. This forms the basis for assessing whether the ecological status of a water body is in accordance with the predefined objectives. The specific objectives and requirements of monitoring in the WFD are described in a guidance document (EC, 2003a). Here some general guidance are provided on where, what and when to monitor. These recommendations do not, however, specify the levels of precision and confidence required by the monitoring programmes, but instead state a key principle: "*the actual precision and confidence levels achieved should enable meaningful assessments of status in time and space to be made. Member States will have to quote these levels in River Basin Management Plans and will thus be open to scrutiny and comment by others*". The acceptable level of precision and confidence is thus a subjective quantity that depends on the socio-economic interests that are at stake and of the risk strategy of the decision makers. In this context it should be borne in mind that the cost of measures to improve the water status are orders of magnitude greater than the costs of monitoring. In general, the lower the desired risk of misclassification, the more monitoring (and hence costs) is required. Therefore, there is a balance between the costs of monitoring against the risk of water bodies being misclassified.

3.2. Modelling categories

Numerous model systems have been developed in the past tailored to describe different parts of the hydrological system in different details. Dependent on the model approach the data requirements vary greatly both with respect to the need for time-series and system data, like soil maps and land use, for more discussion on data availability and accessibility see Refsgaard *et al.* (2010). Here we differentiate between four different model complexities, which may be applicable for different model purposes:

- *Simple models.* These models rely on relatively simple process descriptions, which may not be based on physical/chemical laws, but on an empirical or statistical cause-effect relation. The models may not necessarily be simple to construct or understand, but often they are quick and easy to run. Several such models have been developed within a GIS interface and are generally only distributed spatially and assume all quantities to be temporally invariant.
- *Intermediate models.* Models in this category are normally based on a conceptual understanding of the system and processes. The descriptions are, however, lumped where all parameters and variables represent average values. Hence, the equations of the lumped models are semi-empirical, but still with a physical basis.
- *Comprehensive models.* These models rely on physical laws, and the flows of water and energy are calculated directly from the governing continuum (partial differential) equations. By a distributed description in time and space, these models provide a more detailed description of the hydrological system. Furthermore, models in this category are able to make use of all available data from the system, and provide estimates of all internal state variables, not only those of primary focus.
- *Process studies.* These models are used to validate or refine the underlying relations/equations, which require a much more detailed description to study the relevant processes thoroughly and much more data are required. This type of model is not common in praxis but most often used in the research world by use of research models, where alternative formulations may be tested and compared to field observations from dense monitoring programmes designed for the specific purpose.

The different model concepts provide different degrees of insight in the physical system and have different requirements with respect to system data and data for model calibration and validation. While the empirical models only describe an observed relation between the variables of interest, the physical-based models can, theoretically, describe all state variables of the entire model domain, and thus hold the potential of providing the most detailed and correct representation of the physical system. Most distributed models are, however, a combination of distributed and lumped descriptions, where less important areas/ processes are described by a lumped approach, which lowers the data requirements for the model application.

It must be emphasised that this classification is schematic and that many model codes or applications do not fit exactly. Some model codes fall naturally into one of the above four boxes, while others may be placed in more than one.

The modelling domain is subdivided into the following six groups:

- *Water resources assessment* is the determination of the availability of surface water and groundwater in space and time, including the impacts of the various water uses. Model types include catchment scale models for water allocation, rainfall-runoff, groundwater and integrated/coupled versions of these.
- *Floods* is here concerned with modelling for assessing extents and risks of flooding and for assessing flood control schemes. These models are typically 1D/2D hydraulic/hydrodynamic models of rivers and flood plains, maybe with rainfall-runoff models for providing inflow hydrographs.
- *Surface water quality and ecosystems* is concerned with studies of water quality and ecological aspects in rivers, lakes and estuaries.
- *Agricultural management, including non-point pollution* often deals with field scale studies of irrigation, fertiliser application and leaching and pesticide application and leaching. The most comprehensive of these models also includes simulation of crop yield as a consequence of the applied agricultural management practise.
- *Groundwater quality* is the study of groundwater vulnerability to pollution and the transport and fate of contaminants through the aquifer systems.
- *Socio-economic cost assessment* is the social and economic assessments of various actions considered to improve the ecological condition towards achieving good water status.

Some examples of different model systems appropriate for the different modelling domains are provided in Table 3.1.

Table 3.1 Types of model codes and examples of codes suitable for different domains and different levels of model complexity

Complexity of code *(Data Requirements)*	Water Resources Assessment	Floods	Surface water quality and ecosystems	Agricultural management, including non-point pollution	Groundwater quality	Socio-economic cost assessment
Simple models *(Low)*	GIS based water balance tools, MIKE BASIN, HEC-HMS	Cascade Models (Discrete Linear Cascade Model)	Regression models	Regression models, Export coefficient models	Overlay/index methods (DRASTIC, SINTAC, etc.)	Cost effectiveness
Intermediate models – *(Intermediate)*	Sacramento, HBV, MIKE 11		DYNDIS	USLE, HBV-N, SENSMOD		Hedonic pricing method, Contingent valuation method, Cost-benefit
Comprehensive models – *(High)*	MIKE SHE, MODFLOW	MIKE 11, SOBEK, ISIS HEC-RAS	DELFT3D-ECO	SWAT, DAISY, SOIL-N, ANIMO	PHREAQC, HST3D, MODFLOW, MIKE SHE	
Process studies – *(Very high)*	Comprehensive models + tailor made research codes	Comprehensive models + tailor made research codes	Comprehensive models + tailor made research codes	Comprehensive models + tailor made research codes	Comprehensive models + tailor made research codes	Comprehensive models + tailor made research codes

In general, the more complex model systems may provide more detailed results and thus be used in studies where the discrimination between different processes is critical. However, the more complex the model system becomes, the more data is usually required and more parameter values have to be estimated. It is not possible to provide a finite relation between model complexity, data availability and the model performance. One common illustration of the conceptual relationship is, however, shown in Figure 3.3. For a given model complexity the predictive performance increases with data availability up to a certain limit, beyond which more data do not improve the predictive performance. This limit is reached first for the less complex models, and only the very complex models are able to make full use of large data availability. It also illustrates, however, that for limited data availability the moderately complex models may often perform better than the very complex models. The reason for this is that the more complex models may be able to describe the physical system more correctly, as the simple models rely on more assumptions and thus have a higher uncertainty in the model structure, i.e., incorrect representation of the system. On the other hand, the more complex models do generally require more parameters. If these parameters cannot be measured or are otherwise inaccessible, they must be estimated, which introduce higher uncertainty due to the parameter uncertainty. Thus, for a given data availability the advantages gained in using the more complex models may be counteracted by the increase in parameter uncertainty.

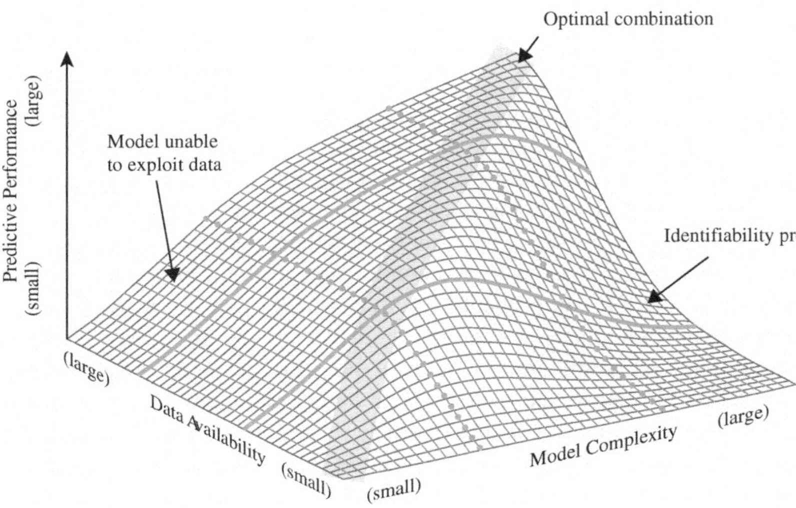

Figure 3.3 Schematic diagram of the relationship between model complexity, data availability and predictive performance (modified from Grayson and Blöschl, 2000)

The view illustrated in Figure 3.3 originates from statistical theoretical considerations assuming that the number of free parameters requiring calibration increases with model complexity. This is, however, not always the case. For model codes with comprehensive application records there are often experience based recommendations for setting default parameter values, so that typically only a handful of parameters are adjusted by calibration in the particular catchment. In this way secondary data sources (all previous model applications in other locations) are used. This would correspond to moving left on the data availability axis towards larger data availability.

3.3. Balancing precision, data and model concept

In the monitoring guidance document (EC, 2003a) some general guidance is provided on where, what and when to monitor in the different monitoring programmes. Most detailed information is provided on which variables must be included in the monitoring programmes, while the required temporal and spatial resolutions in general are linked to the chosen level of confidence and precision, where a key principle is stated: "*the actual precision and confidence levels achieved should enable meaningful assessments of status in time and space to be made. Member States will have to quote these levels in River Basin Management Plans and will thus be open to scrutiny and comment by others*". The acceptable level of precision and confidence is thus a subjective quantity that depends on the socio-economic interests that are at stake and of the risk strategy of the decision makers.

An optimal model approach to support the monitoring obligation does thus depend on both technical aspects, such as type of problem to solve and the problem/system complexity as well as the required confidence. Simple models are therefore likely to be appropriate in a first step evaluation, where all existing data are combined to obtain a first overview of possible areas at risk. But, as the uncertainty in these models are high, they do generally not provide the sufficient basis for specific action strategies, such as detailed design or redesign of monitoring networks. In a socio-economic context the formulation of the programmes of measures is probably the most challenging, as it may have a direct impact on individuals. Here the confidence in the decision making is likely to be high and more complex model systems are needed. More guidance on how to select an appropriate model system based on the problem at hand can be found in the Benchmark Models for Water Framework Directive (BMW) project (http://www.rbm-toolbox.net/bmw/index.php, more discussion on the data availability issue is provided by Refsgaard *et al.* (2010).

4. JOINT USE OF MONITORING AND MODELLING

Although the joint use of modelling and monitoring has not been common praxis in the implementation of the WFD to date, the literature is abundant with case studies where the use of models has contributed significantly in the analysis of the monitoring data, and provided information not easily extractable from the monitoring data alone. Based on such studies five tasks may be identified, where modelling has proved beneficial for objectives similar to those outlined for the WFD monitoring programmes, Figures 3.1 and 3.2:

(1) Quality assurance.
(2) Interpolation in time and space.
(3) Conceptual understanding (Annex II analyses).
(4) Anthropogenic activity.
(5) Design or redesign of monitoring programmes (quantitative/surveillance/operational/investigative).

Table 4.1 provides an overview on how these five tasks relate to the different purposes of the monitoring programmes. For these tasks several operational methods have been developed and the use of models to support the tasks is therefore relatively straight forward and can be included on a routinely basis. A more detailed discussion of the tasks is given in the next section.

In addition to the operational methods which may readily be used for the five tasks listed above, the research community continuously develops methods to improve the joint use of monitoring and modelling. While many of these methods are still under development or too complicated to use in standard model projects, they are likely to be operational in near future and expected to be valuable in a WFD context. A few examples of such methods are provided in Section 4.2.

Table 4.1 Overview of how the described model uses support the different WFD monitoring requirements. Requirements are listed for monitoring related to groundwater (italic) and surface water (bold) and both (italic and bold)

	Quality assurance	Interpolation	Conceptual model	Anthropogenic activities	Design of monitoring programmes
Level Monitoring network			*GW-SW interaction*	*Long term impact by abstraction*	
			Characterisation & Risk assessment, quantitative status	*GW abstractions impact on SW*	
				Saltwater or other intrusions	
Surveillance Monitoring	General Purposes	General Purposes	*Characterisation & Risk assessment, qualitative status*	***Assessment of long-term changes: natural conditions and anthropogenic activities***	***Design of future monitoring programmes***
			Status of GW bodies not at risk		
			Impact assessment		
Operational Monitoring			***Status of water bodies at risk***	***Evaluate effects of POMs***	
				Detection of pollution trends in GW	
Investigative Monitoring			**Pressures unknown (revise conceptual model)**		
			WB at risk but no operational monitoring established		

4.1. Operational modelling support for monitoring

4.1.1. Quality assurance

Description

High quality data is a prerequisite to a sound management of the water resources. Quality assurance is specifically mentioned as an important activity in the WFD guidance document on monitoring (EC, 2003a). However, quality assurance is here restricted to sampling and laboratory procedures. While this is a very important step it is not sufficient to ensure a good data quality. Beside the laboratory analysis, data may be erroneous due to many sources, such as bias or drift in the monitoring equipment, lack of representativeness of the data, malpractice in sampling, erroneous storing of data in databases etc.

A common approach to evaluate monitoring data is by graphical analyses, either constructing time-series or 2D/3D plots, and identify possibly "outliers" that are likely to be erroneous. While this may capture obvious errors it may not always be easy to discriminate between errors and system response. The use of models, describing the system responses to input stresses, may here have provided a much better basis for the detection of errors, Figure 4.1.

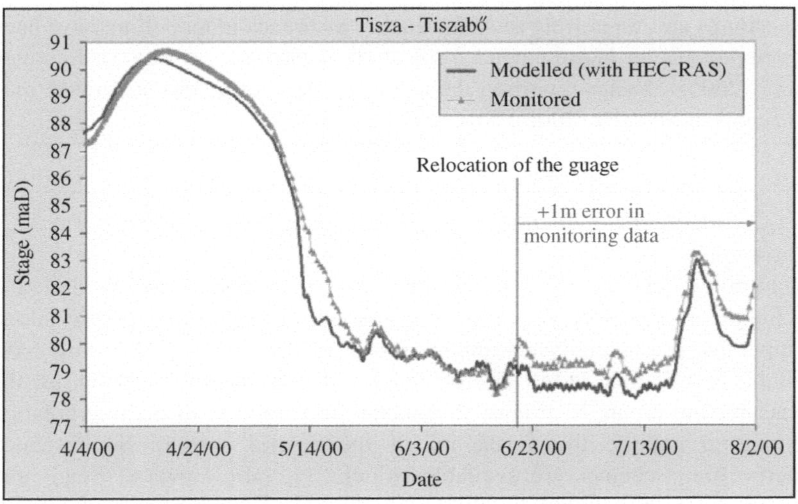

Figure 4.1 Illustration on how a model has been used to identify a relocation of a gauging station, which is not revealed from the monitoring data alone

Similarly, if a monitoring device is located wrong, e.g., expected to be in the vadose zone but actually placed in the saturated zone, or there is a drift or bias in the monitoring device, it may not be detectable from graphical plots. Yet another limitation in simple data analysis is that different data types are evaluated independently and does thus not consider possible linkages between the data types, based on physical/chemical laws. Data therefore need to be used actively to identify possibly errors (besides laboratory analysis) and evaluated if the data are representative.

Methodologies
May often be based on simple graphical plots comparing model results and observed data from which suspicious spatial/temporal regions are identified for further analyses.

Relevance for monitoring
Quality assurance of data by use of models is highly relevant and recommendable for all types of monitoring. This can be seen as a basic analysis of monitoring data and a supplement to the quality assurance on sampling and laboratory procedures described in EC (2003a).

References
Procedures and supporting tools for quality assurance in modelling have been developed in the HarmoniQuA project: Refsgaard *et al.* (2005); Scholten *et al.* (2005). Studies reporting the use of models for quality assurance may be found in Styczen (2002); Styczen *et al.* (2004).

4.1.2. Interpolation and extrapolation in time and space

Description
Monitoring data are discrete in time and space and some interpolation technique is therefore required to transform these discrete points into a temporally and spatially continuous image of the variables. We thus have reliable observations at few discrete points, whereas the reliability outside the measurement points is strongly dependent on the ability of the interpolation routine to capture the variability of the natural system. Sophisticated interpolation schemes are available, which, e.g., allow overall trends and anisotropy to be included, but they will always be limited by their generality and underlying assumptions. In contrast interpolation by models has the ability to include the system characteristics and stresses and is, as such, based

on physical quantities. This is a very important advantage, especially in spatial and/or temporal heterogeneous systems, where the sampling resolution is insufficient to resolve the heterogeneity. An example is monitoring of river systems which are very dynamic and heavy rainfall of short duration (possibly shorter than an hour) results in a sudden change in the discharge. If data is measured at a lower temporal resolution an estimated discharge based on simple interpolation techniques may be significantly in error. Spatial interpolation is important in groundwater management, but generalised interpolation routines do not consider non-stationarity (an aquifer may be homogenous in some parts and very heterogeneous in other parts) but this may be built into the model. The models therefore provide a much stronger interpolation routine, which, in addition, can be improved by adjusting the conceptual model on the basis of new data from field observations.

Extrapolation is specifically relevant in the WFD to establish reference conditions in surface waters. Where possible this reference condition should be based on data from a reference site network, i.e., sites that are unaffected by anthropogenic activities. Where such reference sites do not exist the use of models is often the only available option to extrapolate reference conditions, and the WFD explicitly mentions modelling as a suitable method (EC, 2000) for this purpose.

Methodologies

Interpolation and extrapolation are the actual results from model simulations and predictions. In distributed models spatial interpolation is obtained directly in the resolution used in the model. Temporal interpolation may be discrete or near continuous depending on the model approach (stationary or dynamic), and temporal extrapolation is often accomplished by the formulation of scenarios describing different possible changes in external stresses.

Relevance for monitoring

Model based interpolation of monitoring data is highly relevant and recommendable for all types of monitoring. This can be seen as a basic analysis of monitoring data

References

Nilsen *et al.* (2003) provide an example on extrapolation of reference conditions.

4.1.3. Conceptual model

Description

Monitoring data alone provide only a specific measure like a chemical concentration, river discharge or a groundwater level. This information is, however, of little value for the formulation of management strategies without a thorough understanding of the system with respect to its dynamics and the cause-effect relations. A conceptual model is therefore needed not only to interpret the monitoring data but also to formulation of management strategies. Construction of a conceptual model is therefore a very important but also often a very complicated task, due to the complexity and heterogeneity of physical systems.

A thorough examination of the conceptual understanding must be based on all relevant data types as all data contain (different) information of the same system, and the conceptual model must thus be consistent with all data. In this context, models may be thought of as sophisticated databases that provide an ordered way to store field data in which the relationships between data are defined so that the consistency of the conceptual model can be checked. Confronting the model results with monitoring data may result in either an increased reliability in the conceptual model, or a falsification hereof. In the latter case the monitoring data reveal new information on the system that is not consistent with the present understanding, and the conceptual model must be revised. More monitoring data are then usually required to reformulate the conceptual model.

An erroneous conceptual model may have a crucial impact on the management strategies by misclassification of the water bodies, by the design of ineffective programmes of measures and by non-optimal design of monitoring programmes. A conceptual model will, however, always be a simplified version of the nature and the conceptual model is therefore inherently uncertain. Risks assessment is part of an optimal water management and models may be used in this context to assess the importance of being wrong on the conceptual model. This can be achieved by the formulation of multiple plausible conceptual understandings, which are implemented in the model followed by model simulations and an analysis of the model results.

Methodologies

Test of a conceptual model is often addressed in several stages of a model study, initially at the model construction and later, if applicable for the model system, during the calibration and validation of the model.

Relevance for monitoring

A thorough conceptual understanding of the physical system is vital to all monitoring programmes and is explicitly required in the WFD for the characterisation of the water districts and the assessment of the status and possibly grouping of water bodies. But a conceptual understanding is furthermore a prerequisite for the required risk and impact assessments. The conceptual model must be tested and validated by confronting the model with new monitoring data. If the monitoring data reveal surprises not foreseen by the conceptual model, the model must be revised and it may often be required to make a more comprehensive dedicated monitoring for this purpose, namely investigative monitoring.

References

Neumann and Wierenga (2003); Troldborg (2004); Højberg and Refsgaard (2005).

4.1.4. Assess effects of anthropogenic activities

Description

Monitoring data are used to identify trends in the ecological status and to assess whether an implemented programme of measures has had the expected effects. In this regard the natural climate variability induces variability in the variables to be analysed. Natural variability, together with measurement errors, will act as noise that may hide the signals from and make it very difficult to identify the effects of the anthropogenic activities. This is especially problematic in cases where the full effect of the intervention is slow, and the changes are modest initially and small compared to natural variations. Models may here be useful, if they are able to explain some of the natural variability and thereby enhance the signal from the anthropogenic activity. The key idea is to focus on the differences (residuals) between the model results and the observation data. If the model is able to describe the natural variations the residual does not include the part of the natural variability that the model is able to explain and the variability in the residuals will therefore be less than the variability in the raw observations. A reduced variance in the test series strengthens the tests and makes it easier to identify the signal from the anthropogenic activities. Tests can be made both for (linear) trends and for sudden shifts in conditions.

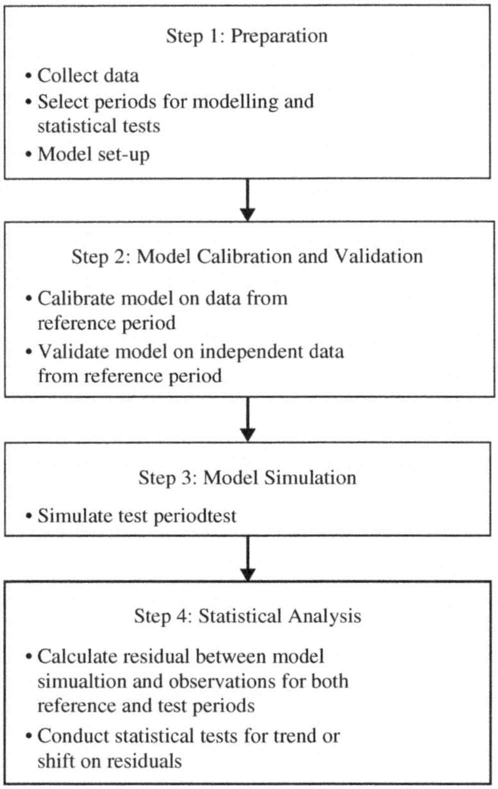

Figure 4.2 Outline of methodology for identifying effects of human activity by using a model to explain some of the climate induced natural variability. (Adapted from Refsgaard *et al.*, 1989)

This approach can be used to filter out both temporal and spatial natural variations, see also Section 4.1.5 below and case study 7.4.

Methodologies
Statistical tests on residuals, e.g.: student t-test and a Wilcoxon Rank-Sum test.

Relevance for monitoring
Identification of trends is a key objective of surveillance monitoring. Furthermore, analysis of the effects of implemented programmes of measures is a key objective of operational monitoring. In both cases it may be difficult to distinguish between the effects of anthropogenic activity and climate variation when analyses are made on the raw observation data alone. As illustrated in the examples above, this can be done much more efficiently by joint use of modelling and monitoring data.

References
Refsgaard *et al.* (1989); Lørup *et al.* (1998); Grath *et al.* (2001).

4.1.5. Design of monitoring programme

Description
There are many different objectives to design and operate a monitoring system. Because the operation of monitoring programmes is expensive, the question is how we can assess the balance between the information content and the cost of the monitoring programme. State variables (pressure, fluxes, concentration, etc.) show variability is space and time. This variability appears as noise, when estimating the characteristics of the state variables. Consequently, estimates of the characteristics are uncertain. In principle the more variability in the state variables, the more measurements are required to arrive at the same level of uncertainty. For the design of a monitoring programme we have to assess the relation between the measurement effort (number of locations and measurement frequency) and the reduction of the uncertainty.

A basic principle in the application of model-based monitoring design is the link of model simulations to optimisation schemes by which the best monitoring strategy under specified constraints can be identified. Commonly, the purpose of the optimisation is to make the monitoring programme cost-effective, but the optimisation may have multiple objectives. The advantage of a model-based design is that the complexity of the physical system can be taken into account in the evaluation of alternative monitoring designs. In particular, physically based models that give a spatially continuous prediction of the state variables, account for a part of the natural variability. Subtracting the model predictions and observation data the part of the natural variability, described by the model, is filtered out from the observations. The noise, or uncertainty, in the monitoring data is consequently reduced, and the monitoring programme can be designed according to the reduced uncertainty. The better the model describes the natural variability, the less monitoring data points are required to arrive at the same level of uncertainty. In the design of an optimal monitoring network, a model, combined with appropriate techniques, can be used to evaluate the information a specific observation holds on the natural system. This can be done for any existing or imaginary sampling point or data type. The model can thus be used to define the number and location of sampling points required to achieve a certain level of uncertainty. Conversely, the model can be used to quantify how much each potential observation will reduce the uncertainty in the model predictions.

Methodologies
The analyses are based on the residuals (observation data minus model predictions) which, due to the models ability to describe the natural variability, have a smaller variance than the observations. With a reduced variance fewer points are required to obtain the same accuracy in an interpolation scheme. Kriging may, e.g., be chosen for interpolation, which can provide an estimate for the accuracy of the interpolation (standard deviation), and the monitoring network can be designed to support a predefined accuracy in the Kriging interpolation. Others approaches are based on filter techniques like Kalman Filtering or representers.

Relevance for monitoring
Optimal design of monitoring network is relevant for all monitoring programmes. Specific for the surveillance monitoring is that data acquired here should provide information for an efficient and effective design of future monitoring programmes.

References
Reed and Minsker (2004); Snepvangers *et al.* (2005).

4.2. Perspectives for future modelling support for monitoring

As previously stated, the link between data and modelling in practical applications is at present a one way relation – data to model, after which the model is run to study a specific problem. A continuous update of the model by utilisation of new monitoring data, commonly referred to as data assimilation, is presently almost restricted to meteorological and flood forecasting applications. But this approach may similarly be used for other domains and has received a growing attention in the research community. Especially the potential of remote sensing has been studied, due to high temporal resolution of satellite images and relatively low cost, but traditional monitoring data have also been applied in data assimilation. Besides the potentials of data assimilation, the joint use of modelling and monitoring is beneficial in an early stage, where the monitoring programmes are designed. This is elaborated above where models calibrated to a specific system can be utilised to optimise the monitoring programme. Models may, however, also be applied in the design of monitoring programmes even before sufficient data are available for calibration, which is accomplished in, e.g., optimal experimental designs, where synthetic model produced observations are employed in the design phase.

Good examples on integration of monitoring and models do exists – as the early warning systems for floods and accidents with chemicals spills as developed in the Rhine and in the Tisza River, but are generally not common in praxis. Below short descriptions are provided for a few methods where monitoring and modelling are closely linked. This illustrates some future perspectives of joint modelling and modelling. The descriptions are only provided as examples and should not be considered a comprehensive list.

4.2.1. Model development

The WFD brings two important challenges: (1) water must be managed in an integrated way, i.e., including all uses of water and all domains, such as surface- and groundwater, and (2) the qualitative status is related not only to the chemical substances measured in the water, but should also be evaluated based on the ecological status. The integration of all domains and the assessments of the chemical and ecological status pose major challenges to both monitoring designing and praxis, see e.g., the Chemical Monitoring Activity (Quevauviller, 2006) and A combined monitoring-based and modelling-based priority setting (COMMPS) (http://ec.europa.eu/environment/water/water-framework/preparation_priority_list.htm), as well as to modelling, e.g., Wasson *et al.* (2003).

In present years much effort is put in developing new models and methods to deal especially with the integrated and ecological aspects. In this regard, much is gained if monitoring and modelling tools and methodologies are developed with a joint application in mind. Most of the obstacles in joint use of monitoring and modelling listed in Section 2.2 can be traced back to the parallel development of monitoring and modelling praxis, with separate focuses. To avoid a similar situation in the future, it is essential that there is a strong collaboration between the two communities, to learn the requirements, advantages and limitations in monitoring and modelling.

4.2.2. Uncertainty reduction in models

Monitoring systems are generally not designed to provide information to design or calibrate models. Therefore, often the data are not available from locations and times that are optimal for the model. The available data influence the models in two ways:

- *Parameterisation of the model.* The detail of the model is restricted by the availability of data. In areas with little data, the model cannot have much detail. The number of independent variables has to be in accordance with the data available. In the ideal case, if a detailed model is required, the monitoring system should be designed to support a detailed parameterisation.
- *Calibration of the model.* Often measurements are not taken at locations and times that are most sensitive to the model performance. Therefore, from the model prediction point of view, the monitoring is not the most cost effective. Reallocation of the monitoring points can reduce the model uncertainty, while the cost of monitoring remains the same.

Reduction of model uncertainty can be achieved by quantifying the relation between the uncertainty in the model and the monitoring system. Not only is the impact of existing monitoring locations on the uncertainty reduction considered, but also the impact of potential locations. Monitoring locations and measurement times can then be ranked, according to the impact on the model uncertainty. The monitoring locations and measurement times are selected based on the highest impact on the reduction of uncertainty. So, the design criterion for the monitoring system is the model uncertainty.

4.2.3. Flood forecasting

Flood forecasting tools are driven by hydrological model systems that calculate future discharges and water levels by routing runoff from the sites of precipitation events, through the different runoff paths (surface runoff, subsurface runoff and channel flow), down to the sections where forecasts are demanded. Boundary conditions for forecasting consist of measured and forecasted hydro-meteorological data coming from gauging stations as well as from (other) forecasting systems. Flood forecasting tools are operated on continuous or event basis using real-time hydro-meteorological boundary data. Future discharges and water levels are calculated at each forecasting station with an appropriate lead time.

Flood forecasting systems fully integrate monitoring and modelling by use of data assimilation. This implies that the monitoring data are used in real-time to update (i.e., correct) the hydrological model.

5. DISCUSSION AND CONCLUSIONS

Monitoring data have always been requested by modellers to convert a generic model into a site-specific model. Modellers have thus long acknowledged the worth of monitoring data and been concerned in how monitoring programmes can be designed to improve the data foundation for modelling. Monitoring people have, on the other hand, not been dependent on models to design, implement and complete a monitoring programme. Traditionally, a good monitoring data set has been equalled long time-series, and the use of models to interpret and evaluate a monitoring programme is rarely seen in practice. While modelling may not necessarily be required for all monitoring activities at any site-specific location, the methods and examples presented here clearly demonstrate the potentials of using models to utilise the monitoring data in a much more efficient way.

The present Guidance Report focuses only on the combined use of modelling and monitoring. Potential use of models in the WFD is, however, not limited to the monitoring obligations. As described in Hattermann and Kundzewicz (2009), models may be beneficial and – in a variety of circumstances – indispensable when implementing the WFD, e.g., for impact prediction and what-if analysis. They may help to understand and optimise the use of resources and can in particular be useful in the following steps of the WFD, Figure 5.1.

- Assessment of the currents status and preliminary gap analysis
- Establishment of monitoring programmes
- Gap analysis
- Setting up the programmes of measure
- Development of river basin management plans
- Implementation of the programmes of measures and preparation of the interim report on the implementation
- Information and consultation of the public

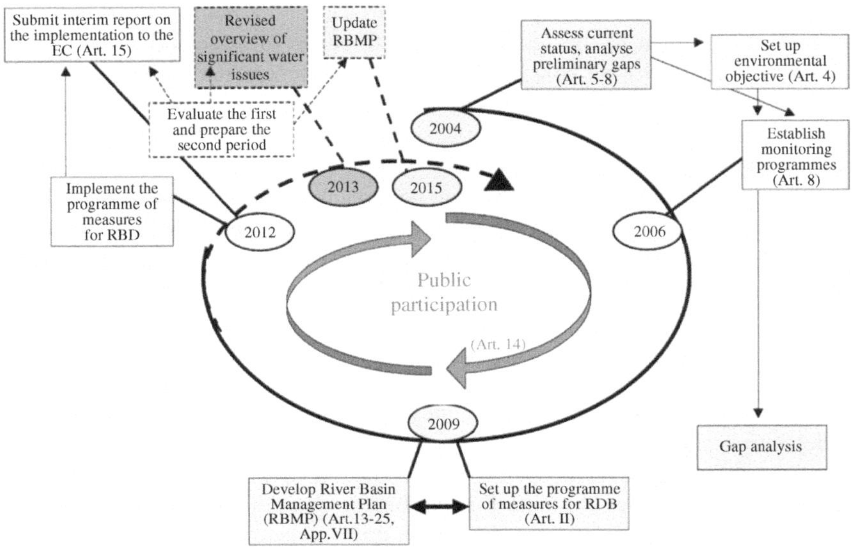

Figure 5.1 The WFD cycle displaying the sequence and relation between the WFD activities

The Harmoni-CA workshops on joint use of modelling and monitoring showed that most often models are not considered an option when the monitoring obligations in the WFD are solved in practise. A number of obstacles were identified to why this is so. Most of these obstacles are settled in historical, cultural and institutional aspects, but the workshops illustrated that if people from the monitoring community and from the modelling community are brought together for open-minded discussions and with free hands they are very open to discuss ideas and suggestions from the other side of the table, and able to see the perspectives in the combined use. To advance the joint use of monitoring and modelling there appears to be a strong need to first of all raise the awareness of how monitoring and modelling can be combined. Additionally, a shift in the perception of models is needed, where currently models are often considered useful only to provide answers to specific questions. This view must be replaced by recognition of models as being valuable tools that should be used continuously and interactively along with monitoring data, a tool that serves many purposes from the design phase of monitoring programmes to the interpretation and quality assurance of the monitoring data.

The overall objective of the Water Framework Directive is to achieve a good quantitative and qualitative status of all waters by 2015. The WFD brings two important challenges: (1) water must be managed in an integrated way, thus including all uses of water and all domains, such as surface- and groundwater, and (2) the qualitative status is related not only to the chemical substances measured in the water, but should also be evaluated based on the ecological status. To fulfil these requirements development of new tools and models is a necessity. To avoid a future parallel development of monitoring and modelling praxis in the field of integrated and ecological assessments, it is of vital importance that monitoring and modelling are realised to be inter-linked activities and not independent disciplines.

6. REFERENCES

Arustiene, J., Vaitiekuniene, J. and Jørgensen, L.F. (Eds.) (2005). *Joint use of modelling and monitoring for implementing the Water Framework Directive. Potential Modelling Support*. Workshop report, Harmoni-CA WP2 & WP4. D4-4. Available at http://www.harmoni-ca.info/

Borowski, I. and Hare, M. (2007). Exploring the gap between water managers and researchers: Difficulties of model-based tools to support practical water management, *Water Resources Management* **21**, 1049–1074.

Brugnach, M., Tagg, A., Keil, F. and de Lange, W.J. (2007). Uncertainty matters: Computer models at the science–policy interface. *Water Resources Management* **21**, 1075–1090.

EC (2000) Directive 2000/60/EC of the European Parliament and of the Council of October 23 2000 establishing a framework for Community action in the field of water policy. *Official Journal of the European Communities*, L327/1–L327/72, 22.12.2000.

EC (2003a). *Water Framework Directive, Common Implementation Strategy, Working Group 2.7. Monitoring*. Available on http://forum.europa.eu.int/Public/irc/env/wfd/library

EC (2003b). *Water Framework Directive, Common Implementation Strategy, Working Group 2.9. Planning processes*. Available at http://forum.europa.eu.int/Public/irc/env/wfd/library

Grath, J., Scheidleder, A., Uhlig, S., Weber, K., Kralik, M., Keimel, T. and Gruber, D. (2001). *The EU Water Framework Directive: Statistical aspects of the identification of groundwater pollution trends, and aggregation of monitoring results*. Final Report. Austrian Federal Ministry of Agriculture and Forestry, Environment and Water Management (Ref.: 41.046/01-IV1/00 and GZ 16 2500/2-I/6/00), European Commission (Grant Agreement Ref.: Subv 99/130794). Vienna, Austria.

Grayson, R. and Blöschl, G. (2000). Spatial modelling of catchment dynamics. In: Grayson R and Blöschl, G. (Eds.) *Spatial Patterns in Catchment Hydrology: Observations and Modelling*. Chapter 3, 51–81. Cambridge University Press, UK.

Hattermann, F.F. and Kundzewicz, Z.W. (2009). *Model-Supported Implementation of the Water Framework Directive – A Water Manager's Guide*. IWA Publishing, London, UK.

Holt, M.S., Fox, K., Grießbach, E., Johnsen, S., Kinnunen, J., Lecloux, A., Murray-Smith, R., Peterson, D.R., Schröder, R., Silvani, M., ten Berge, W.F.J., Toy, R.J. and Feijtel, T.C.M. (2000). Monitoring, modelling and environmental exposure assessment of industrial chemicals in the aquatic environment. *Chemosphere* **41**, 1799–1808.

Højberg, A.L. and Refsgaard, J.C. (2005). Model uncertainty-parameter uncertainty versus conceptual models. *Water Science and Technology* **52**(6), 177–186.

Højberg, A.L., Refsgaard, J.C., van Geer, F., Jørgensen, L.F. and Zsuffa, I. (2007) Use of models to support the monitoring requirements in the Water Framework Directive. *Water Resources Management* **21**, 1649–1672

Kamphorst, E., Jørgensen, L.F., van Griensven, A. and Vanrolleghem, P.P. (Eds.) (2005). *Joint use of modelling and monitoring for implementing the Water Framework Directive. State of the art on existing monitoring programmes around Europe.* Workshop report, Harmoni-CA WP2 & WP4. D4-3. Available at http://www.harmoni-ca.info/

Lørup, J.K., Refsgaard, J.C. and Mazvimavi, D. (1998). Assessing the effect of land use change on catchment runoff by combined use of statistical tests and hydrological modelling: Case studies from Zimbabwe. *Journal of Hydrology* **205**, 147–163.

Neuman, S.P. and Wierenga, P.J. (2003). *A Comprehensive Strategy of Hydrogeologic Modeling and Uncertainty Analysis for Nuclear Facilities and Sites.* University of Arizona, Report NUREG/CR-6805.

Nilsen, K., Somod, B., Ellegaard, C. and Krause-Jensen, D. (2003). Assessing reference conditions according to the European Water Framework Directive using modelling and analysis of historical data: An example from Randers fjord, Denmark. *AMBIO* **32**(4), 287–294.

Parr, T.W., Sier, A.R.J., Battarbee, R.W., Mackay, A. and Burgess, J. (2003). Detecting environmental change: Science and society – Perspectives on long-term research and monitoring in the 21st century. *Science of the Total Environment* **310**, 1–8.

Quevauviller, P. (2005). Groundwater monitoring in the context of EU legislation: Reality and integration needs. *Journal of Environmental Monitoring* **7**, 89–102.

Quevauviller, P. (2006). Chemical monitoring activity under the common implementation strategy of the WFD. *Journal of Soils and Sediments* **6**, 2–3.

Reed, P.M. and Minsker, B.S. (2004). Striking the balance: Long-term groundwater monitoring design for conflicting objectives. *Journal of Water Resources Planning and Management – ASCE* **130**, 140–149.

Refsgaard, J.C., Alley, W.M. and Vuglinsky, V.S. (1989). *Methodology for Distinguishing Between Man's Influence and Climatic Effects on the Hydrological Cycle.* IHP-III Project 6.3. Technical Documents in Hydrology, UNESCO.

Refsgaard, J.C., Henriksen, H.J., Harrar, W.G., Scholten, H. and Kassahun, A. (2005). Quality assurance in model based water management – Review of existing practice and outline of new approaches. *Environmental Modelling and Software.* **20**, 1201–1215.

Refsgaard, J.C., Jørgensen, L.F. and Højberg, A.L. (2010). Data availability and accessibility: in view of model-based water management. In: Vanrolleghem, P.A. (Ed.) *Integrated Assessment for WFD implementation: Data, Economic and Human Dimension – Water Framework Directive Series, Volume 2.* IWA Publishing, London, UK. (this volume).

Scholten, H., Kassahun, A., Refsgaard, J.C., Kargas, T., Gavardinas, C. and Beulens, A.J.M. (2005). A methodology to support multidisciplinary model-based water management. *Environmental Modelling and Software* **20**, 2101–1215.

Snepvangers, J.J.J.C., Berendrecht, W.L. and Valstar, J.R. (2005). Model and measument based optimisation of quantitative monitoring networks: A comparative study. Modelcare 2005. *Prepublished proceedings* 145–152.

Styczen, M. (2002). Development of a tool for estimation of pesticide occurrence in surface water under Danish conditions. *International Journal of Environmental Analytical Chemistry* **82**, 611–630.

Styczen, M., Petersen, S., Kristensen, M., Jessen, O.Z., Rasmussen, D., Andersen, M.B. and Sørensen, P.B. (2004). *Calibration of Models Describing Pesticide Fate and Transport in Lillebæk and Odder Bæk Catchment.* Danish Environmental Protection Agency, Pesticide Research no 62, pp. 217.

Troldborg, L. (2004). *The influence of conceptual geological models in the simulation of flow and transport in Quaternary aquifer systems*, PhD. thesis, Geological Survey of Denmark and Greenland Report 2004/107, pp. 120.

Wasson, J.G., Tusseau-Vuillemin, M.H., Andréassian, V., Perrin, C., Fauer, J.B., Barreteau, O., Bousquet, M. and Chastan, B. (2003). What kind of water models are needed for the implementation of the European Water Framework Directive? Examples from France. *International Journal on River Basin Management* **1**, 125–135.

Zsuffa, I., Pataki, B. (2005). *Memorandum: WFD related monitoring and modelling. Suggestions for mutual use and benefits*. Harmoni-CA WP4. Available at http://www.harmoni-ca.info/

7. APPENDIX – CASE STUDIES

In this section a few examples on the joint use of monitoring and modelling are provided for different domains and with different purposes of the study. The examples includes some of the tasks described in the previous sections as indicated in Table 7.1.

Table 7.1 Overview of the case studies and the tasks included in the studies

Task/Case	DK-model	Matsalu river	NHFS	Monitoring design
Quality assurance	X		X	
Interpolation/Extrapolation	X	X	X	X
Conceptual model	X		X	
Anthropogenic effects	X	X		
Design of monitoring programmes	(X)	X		X
Data assimilation	(X)		X	

7.1. The national water resources model for Denmark (DK-model)

7.1.1. DK-model structure and purpose

The quality and quantity of the groundwater has long been of major concern in Denmark where groundwater constitutes 99% of the drinking water supply. To improve the quantitative estimate of the exploitable groundwater resource a national water resource model (DK-model) was established, covering the entire 43,000 km^2 of Denmark (Henriksen *et al.*, 2003). The experiences gained from the work formed the basis for specifying modelling to become a supportive element in the coming WFD monitoring in Denmark. The aspects to which modelling contributes are illustrated below.

7.1.2. DK-model and monitoring

Quality assurance

Climatic data, i.e., precipitation and evaporation, were obtained from two national research institutes, Danish Meteorological Institute and Danish Institute of Agricultural Sciences, respectively. Both institutes have extensive quality assurance on data sampling and use a scientific based state-of-the-art approach to correct the precipitation data (accounting for wind effects and wetting) and estimate evaporation (modified Penman equation). From a scientific standpoint the data resembled thus the best and most reliable data set. But when data were applied in the national water resources model and model outputs compared to measured groundwater head and river discharge data, significant water balance errors were observed. This error could only be explained by inconsistency in the climatic input data. While the precipitation and evaporation data were sufficiently accurate when only the precipitation or the evaporation was of interest, the methods used were not adequate when the entire water balance was the target. As such the inconsistency was crucial for the assessment of the quantitative status of the water resource, and certainly crucial for the water management at local and regional scales. These findings resulted in the collaboration of four national institutes involved in the assessment of the water resource. It was concluded that the water balance problem could not be fully solved based on the available knowledge, but a consensus was agreed (Plauborg *et al.*, 2002) providing recommendations for the correction of precipitation and estimation of the evaporation, which are now used nationwide for estimation of net precipitation.

Interpolation

By simulating the entire country the model is used effectively as a spatial and temporal interpolation tool. For example, continuous images of the groundwater heads in different geological layers are obtained. This is of interest for identifying the areas where the groundwater head is critically low, such as areas with phreatic conditions in chalk formations, which may result in oxidation and release of nickel. Complete images of the groundwater level are similarly required to quantify the lowering of the water table by abstractions. Besides the groundwater head, the model provides three-dimensional interpolation of the groundwater recharge at different depths, so that the sustainability of the groundwater exploitation can be assessed by comparing the total abstraction in an aquifer to the total net recharge to the aquifer.

Conceptual model

The establishment of the DK-model prompted the construction of a nationwide conceptual model on the hydrological system, including the subsurface geology, overland flow processes, unsaturated flow, saturated flow and groundwater-river interactions. Calibration and validation of the model improved the conceptual understanding in several instances. Examples include the observation of abstraction wells that went dry in the simulations, although the correct abstraction rates were applied. This is a typical indication of an erroneous hydrogeological model where the aquifer containing the abstraction well is misinterpreted and the simulation thus provided feed-back to the geological interpretation at the local scale.

The DK-model provides an initial nationwide conceptual model. Since the model construction, the Danish counties, who act as the regional water authorities, have put much effort in updating the geological and hydrological understanding, which has been required for the regional water management by national legislation. The DK-model has been updated as a joint project between the regional water managers and the Geological Survey of Denmark and Greenland, where especially the conceptual model was revised. The numerical model provides a logical framework for this update, where the present conceptual understanding is confronted with the new knowledge, which is accommodated by the overlay of the two conceptual models and illustrated graphically. Differences in the two conceptual models are then discussed, and the initial conceptual model is adjusted as agreed upon.

The construction of nationwide conceptual model has proved valuable, as the national water resources model has been employed in the revision of groundwater bodies in Denmark, to form a consistent basis for the delineation in large parts of Denmark.

Anthropogenic activities
The overall objective of the DK-model is to provide an estimate of the exploitable resource. An acceptable exploitation is clearly linked to the level of impact on the physical system we accept as a consequence of human activities. Groundwater abstraction is the only anthropogenic activity considered explicitly in the model, and two critical aspects have been identified constraining the acceptable exploitation: (1) induced groundwater recharge to the deeper aquifer as a response to abstraction, this is critical as the younger water are generally polluted, and (2) reduction of discharge in the stream, which lowers the potential for achieving a good ecological status in the stream. The impact by water abstractions have been evaluated in a series of model simulations in which the abstraction rates were varied. Furthermore, the robustness to climate change was assessed by making simulations for dry and wet climate conditions.

Design of monitoring programmes
To date the national water resources model has not been used in a monitoring network design phase. However, the national level monitoring network is to be evaluated and redesigned, and it is the intention that this will be aided by the use of the national water resources model.

Data assimilation
Sophisticated data assimilation is not planned to be included in the national water resource model at the present stage. However, intensive geological characterisation is still ongoing in Denmark and it is the ambitions that the updated (hydro) geological knowledge is utilised to update the national water resource model.

7.1.3. References

Henriksen, H.J., Troldborg, L., Nyegaard, P., Sonnenborg, T.O., Refsgaard, J.C. and Madsen, B. (2003). Methodology for construction, calibration and validation of a national hydrological model for Denmark. *Journal of Hydrology* **280**, 52–71.

Plauborg, F., Refsgaard, J.C., Henriksen, H.J., Blicher-Mathiesen, G. and Kern-Hansen, C. (2002). *Vandbalance på mark- og oplandsskala*. DJF rapport 70. Danmarks JordbrugsForskning, 45 pp. (in Danish).

7.2. Matsalu River Catchment

7.2.1. Modelling area and purpose

Modelling studies are commonly expected to require a vast amount of data to setup and run the models. Such data may not be readily available and thus hinder the use of models without intensive data collection prior to the study. Although the reliability of the model results generally are proportional to the amount of data, valuable model studies may be carried out with less data, which, in turn, may identify locations and data types that should be collected to improve the model and thereby the reliability of the results and future decisions. Such model exercise was undertaken for the Matsalu River Catchment in Estonia, the example is based on the example by Jensen (2005).

The Matsalu River Catchment was studied with the purpose of:

- describing existing water quality
- delineating the water bodies in accordance with WFD obligations
- upgrading the monitoring programme to fulfil the requirement of the WFD
- providing recommendations for pollution abatement to achieve the WFD requirement

The catchment size is 4,870 km^2 and inhabited by 61,315 people (year 2002), of which 28,150 are served by public wastewater treatment plants. The model system MIKE BASIN was used to provide an overview of the anthropogenic load of nutrients (Nitrogen and Phosphorus) and their spatial distribution within the system. The existing monitoring programme included only 7 monitoring stations, Figure 7.1. The stations are primarily located in the downstream part of the river catchment.

7.2.2. Matsalu model

The MIKE BASIN model was setup by utilising standard data such as digital elevation model (DEM) and precipitation. From information on land use (soil types, crops, use of fertilisers, livestock population) the agricultural non-point pollution was estimated for the different parts of the catchment using the GIS facilities of the tool. The transport of pollutants (BOD, Nitrogen and Phosphorus) was estimated taking into account a distance dependant (source to river distance) decay rate in the individual catchments. Pollution input from domestic sources was estimated based on information about human population density, coverage by sewer system and types of treatment plants. Furthermore, other point sources e.g., industrial input were included in the model set-up.

Based on this information, the pollution input from each part of the river system was calculated. The model was calibrated to the best possible extent based on data from the existing seven monitoring stations. As a part of the calibration procedure the turnover of pollutants within the sub-catchment (from source to river) as well as within the river system was adjusted. The result of this calibration process is illustrated in Figure 7.2 with respect to nitrogen and phosphorous. Based on this characterisation water bodies of different quality could be delineated and monitoring stations could be identified with the purpose of following the trend in water quality at the most critical location.

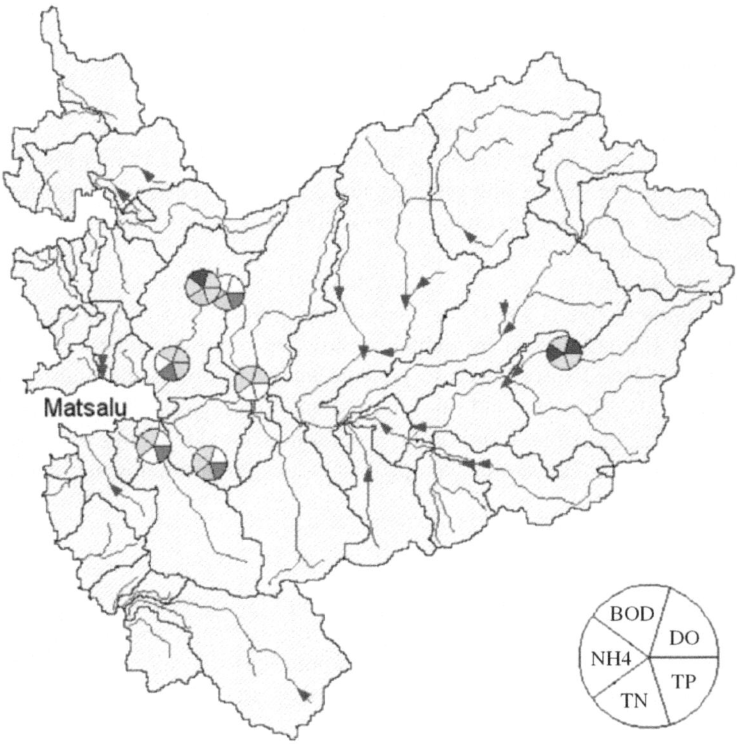

Figure 7.1 Existing monitoring station and water quality classification (1999–2002) of the Matsalu River Catchment (from Jensen, 2005)

The GIS-based modelling tool MIKE BASIN combines GIS-coverages on the relevant characteristics of the system to achieve a combined effect of the physical properties (e.g., river configuration, runoff properties, soil types etc.) and the anthropogenic activities (land use, crops, fertilizer etc.). The main advantages of such modelling approaches are that they generally require less data than the more detailed model systems and are easier to setup and run. However, as for all models the reliability of the results are highly dependent on the quality of the input data and the observations available to test (calibrate and validate) the model, as well as the underlying model assumptions. When data is sparse the model results may be associated by significant uncertainties, and it may not be recommendable to formulate expensive programmes of measures based such model studies. Nevertheless, a model study as presented for the Matsalu River Basin – are very valuable to identify areas that are most problematic which should be examined in further detail.

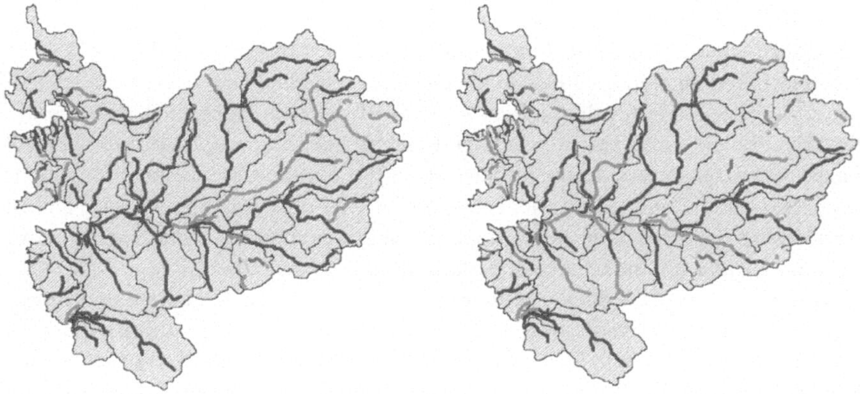

Figure 7.2 Classification of the quality condition in the river systems based on the modelled annual average total N (left) and total P (right) concentration (present condition): Colour code describes the ecological status. For the colour version of this figure please go to http://www.iwawaterwiki.org/xwiki/bin/view/Articles/IntegratedAssessmentforWaterFrameworkDirectiveImplementation (from Jensen, 2005)

7.2.3. Matsalu model and monitoring

In addition to the few monitoring points, data input for the Matsalu catchment was sparse, and to a large extent estimated based on statistics, e.g., land use, live-stock and population densities and location of point sources. By using a model describing the main processes in combination with the available monitoring data, it was possible to test the reliability of the estimated input data, and to interpolate the effect of the pressures for the entire catchment. The final model results depict river stretches and water bodies most likely to meet a good ecological status, or most likely at risk of not meeting the requirements. Using the model in an uncertainty assessment, the reliability of the model results can be addressed, and used to identify areas where more detailed input data and possible more monitoring data and detailed modelling are required to achieve a sufficient level of confidence in the assessment.

7.2.4. References

Jensen, J.K. (2005). Modelling-A tool for interpretation and optimisation of monitoring programmes. In: Arustiene, J., Vaitiekuniene, J. and Jørgensen, L.F. (eds.) *Joint use of modelling and monitoring for implementing the Water Framework Directive*. Workshop report. Harmoni-CA. Available on http://www.harmoni-ca.info/

7.3. National Hydrological Forecasting System of Hungary (NHFS)

Hungary is one of the most flood-exposed countries in Europe. About 1/4 of its territory is endangered by floods of the Danube, Tisza and Drava rivers. In spite of the strong flood control dike systems, states of extreme emergencies frequently develop, which could easily turn into disasters unless appropriate flood control measures are taken. Just during the past decade six extreme, record-breaking floods occured on the Danube and the Tisza, and it was only a matter of luck and of painstaking and efficient flood control measures that no catastrophes have happened, although the chance at certain instances was rather high.

One of the corner-stones of efficient flood control is the accurate and timely forecasting of flood events, which enables the defenders and authorities to take the necessary measures. For this purpose the Environmental Protection and Water Management Research Institute (VITUKI) developed the National Hydrological Forecasting System (NHFS) (Bartha *et al.*, 1983; Szöllősi-Nagy, 1989; Bálint *et al.*, 1996), which has been being operated and periodically upgraded by the institute since 1983. The overall objective of NHFS is to forecast continuously the hydrological regime of major rivers of Hungary with special regard to flood events.

7.3.1. Structure and operation of the NHFS

Forecasts made by NHFS are based on various meteorological and hydrological input data collected real time from the river basins. Measured precipitation, snow, temperature and water level data are collected from upstream gauging stations on daily basis. The frequency of these measurements ranges from hourly, through daily to weekly, depending on the temporal dynamism of variables. Meteorological data are collected from 96 Hungarian and 410 foreign (German, Austrian, Slovakian, etc.) stations, while hydrological data are gathered from 170 Hungarian and 210 foreign stations. Measured meteorological and hydrological time series are extended into the future by appending forecasted data generated by meteorological forecasting systems as well as by hydrological forecasting systems operated by upstream countries.

The collected and assimilated hydro-meteorological data are input to the *forecasting model system* (Bálint *et al.*, 1996; Gauzer and Bartha, 1999), which consists of *rainfall-runoff* and *channel routing* model blocks representing the different sub-catchments and river reaches, Figure 7.3.

Rainfall-runoff models are complex model systems simulating runoff from the mountainous parts of the basins. The following processes are simulated by these models: snow accumulation and ablation, soil frost, rainfall interception, distribution of runoff quantities among the different paths (surface runoff, interflow, base-flow) and finally the actual processes of surface and subsurface runoffs. The different runoff processes are modelled with kinetic wave, continuous cascade or Discrete Linear Cascade (DLCM) models. The two latter are based on the principles of the Nash cascade model.

For routing runoff further in the river channels, DLCM models are applied. Each river reach (reach between two major gauges) are represented by a DLCM cascade (see Figure 7.3). Discharges forecasted by the model system are corrected by an Auto-Regressive Moving Average (ARMA) based statistical error correction technique. Since DLCM models calculate discharges only, water levels have to be derived subsequently with the help of Q-H rating curves. In those river reaches however where no such rating curves are applicable (e.g., reaches upstream of barrages or tributary mouths), a hydraulic backwater model, similar to the one applied by Todini and Bossi (1986), is used for calculating water levels.

The forecasting model is run on daily basis using updated hydro-meteorological data as boundary conditions. Future discharges and water levels are calculated at each forecasting gauging stations (see Figure 7.3) with a lead time of 6 days.

The different modelling modules of NHFS need to be periodically recalibrated and re-verified. For this purpose the NHFS is turned into simulation mode and calibration/validation runs are implemented using historical hydro-meteorological data.

7.3.2. NHFS and monitoring

The primary role of NHFS is hydrological forecasting, which means that monitoring is considered only as a source of real time boundary data. Other important aspects of joint use of monitoring and modelling are taken into consideration only to a limited extent, even though there are considerable potentials in these fields.

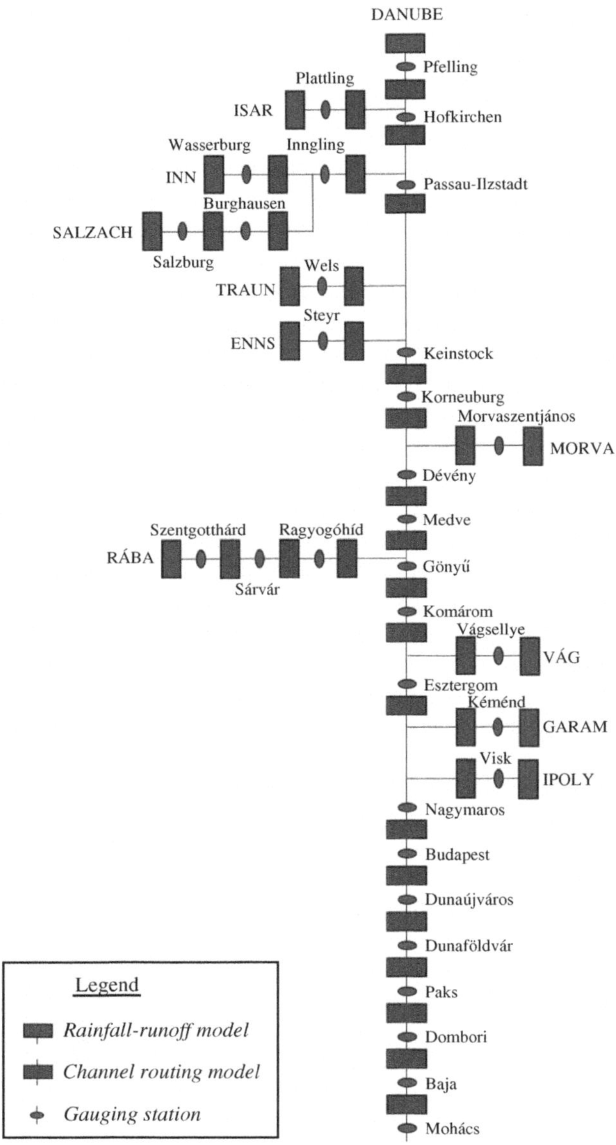

Figure 7.3 Configuration of the NHFS model system for the Danube basin

Quality assurance

For the time being filtering of gross measuring errors is taking place by comparing daily measurements with forecasts from the day before. More accurate error checking could be achieved if measured values were compared with simulated, and not with forecasted values. It is thus recommended to implement regular quality assurance simulations with the help of NHFS.

Interpolation

Running NHFS in simulation mode also enables temporal interpolations between measured water level and derived discharge data. Spatial interpolation on the other hand is not applicable as it is not possible to associate sections within a river reach with components of the cascade model from where calculated discharges can be derived. From this point of view cascade models behaves like lumped models.

Conceptual model

Developing conceptual models and NHFS model systems for the major rivers of Hungary were interlinked, mutually supporting processes. Original concepts, based on historical data and expert experiences, were first implemented and tested in NHFS that provided feedback on how to improve the conceptual model, which in turn formed the basis for developing further the models themselves. This iterative process has been going on since the introduction of NHFS.

Runoff from a river basin can be classified into three general classes: surface runoff, subsurface runoff and river channel runoff. Hydraulic principles of these runoff types are quite different requiring distinct modelling approaches. An essential task of conceptual model development is thus to determine the dominant runoff type(s) in the different subsystems of the basin. Based on such subdivision, models can be associated with the subsystems, from which the complex model system can be built up, see Figure 7.3. Accordingly rainfall-runoff models have been assigned to the mountainous sub-catchments where surface and subsurface runoffs are dominating, while channel routing models have been associated with the lowland parts of the basin where runoff is concentrated in river channels, see Figure 7.4.

With respect to mountainous sub-catchments the development of conceptual model begins with the identification of determining processes such as snow accumulation and ablation, interception, surface runoff, interflow and base-flow. This is followed by the exploration of interactions between these processes. The rainfall-runoff modules of NHFS have been providing continuous and valuable support through testing the different concepts of catchment hydrological processes.

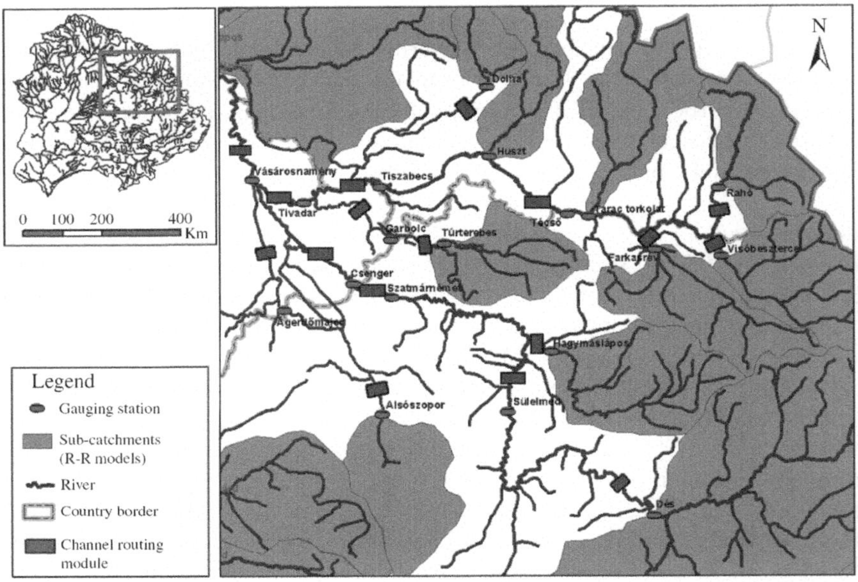

Figure 7.4 Identification of rainfall-runoff sub-catchments and channel systems in the Upper Tisza basin

As indicated above, the conceptual models of the Danube, Tisza and Drava basins, as well as their realizations in NHFS have been continuously improved as gradually more information has been gained (especially during floods) about the behaviour of the systems. The most important milestones of this development were as follows (see also: Gauzer and Bartha, 1999; Bálint *et al.*, 2005):

- An essential feature of DLCM is that the travel time of floods is constant. In reality however travel time depends on the magnitude of discharge and this dependence has proven to be so significant on lowland river reaches that it had to be taken into consideration in the models as well. Accordingly three parallel DLCM cascades, each with different parameters, have been defined for the lowland reaches of the Danube, Tisza and Drava rivers. At low waters only one cascade conveys water, while the other two join in gradually as discharges exceeds the mean and high water thresholds.
- Calculation of water levels upstream of barrages and river mouths has proven to be problematic due to the unambiguous relationships between

discharges and water levels. This led to the development and application of hydraulic backwater models that derive water levels along such reaches as a function of discharges calculated by DLCM.

- Calculating spatial average meteorological input on the sub-catchments has been improved as well. The original method simply calculated the weighted average of data from the gauging stations. Since most stations are situated on lower altitudes, this method failed to take the different meteorological conditions of higher altitudes into consideration. The new method interpolates values at the nodes of a dense grid (the meteorological grid) with the help of a statistical meteorological model that takes topographical elevation into consideration. Sub-catchment average values are than calculated from the grid values.

In spite of the significant model developments there are certain aspects of the state-of-the-art conceptual models that exceed the modelling capacity of NHFS – only fundamental changes in the model system would make it possible to cope with these aspects. One such aspect is downstream influence on discharges, which the DLCM is unable to take into consideration. Downstream influence plays a determining role on the lowland reaches of the River Tisza, where the impounding effects of the Körös and Maros tributaries as well as that of the Danube influence not just water levels but also the discharges of the river. The ultimate solution for this problem would be the replacement of DLCM with a more sophisticated hydraulic model.

Data assimilation
Real time integration of different hydro-meteorological boundary data into a hydrological modelling system makes NHFS an ideal data assimilation framework.

Meteorological data, coming from monitoring stations as well as from meteorological forecasting models, are interpolated and calculated at the nodes of the meteorological grid in order to generate a homogeneous input dataset of appropriate temporal and spatial resolutions (Bálint *et al.*, 2005). For the time being meteorological data from the following sources are assimilated in this way:

- Meteorological stations in Germany, Switzerland, Austria, Czech Republic, Slovakia, Hungary, Italy, Slovenia, Croatia, Ukraine, Romania and Serbia
- ALADIN-HU: meteorological forecasting model of the National Meteorological Service of Hungary (forecasting lead time: 48 h)

- Meteorological forecasting model of Germany's National Meteorological Service (DWD) (lead time: 7 d)
- Meteorological forecasting model of the European Centre for Medium Range Weather Forecasting (ECMWF) (lead time: 10 d)

Similar assimilation is taking place in case of hydrological input data too. Measured water level and derived (from Q-H curves) discharge data from the gauging stations are supplemented with values forecasted by hydrological forecasting systems operated by Germany and Austria. The so assimilated hydrological datasets, along with the assimilated meteorological datasets, form the boundary conditions for NHFS.

Hydrological data assimilation could further be extended with the help of NHFS through repeated temporal interpolations implemented in simulation mode.

7.3.3. References

Bálint, G., Csík, A., Bartha, P., Gauzer, B. and Bonta, I. (2005). Application of meteorological ensembles for Danube flood forecasting and warning. In: *Transboundary Floods: Reducing Risks through Flood Management*, Marsalek, J., Stancalie, G. and Balint, G. (Eds.). NATO Science Series, IV. Earth and Environmental Sciences. Dordrecht, The Netherlands – Vol. 72, 57–68.

Bálint, G., Pándi, G. and Illés, L. (1996). Analyses of floods on the Upper-Tisza with special emphasis on hydrological forecasting. In: Proceedings *18th (XVIII) Conference of the Danube Countries on Hydrological Forecasting and Hydrological Bases of water Management*, Schriftenreihe zur Wasserwirtschaft, Vol. 19/1, Technische Universität Graz, Graz – Austria, August 1996, pp. B77–82.

Bartha, P., Szöllősi-Nagy, A. and Harkányi, K. (1983). Hidrológiai adatgyűjtő és előrejelző rendszer. A Duna. (Hydrological data collection and forecasting system, The Danube), *Vízügyi Közlemények* 65, 3, 373–388. (in Hungarian).

Gauzer, B. and Bartha, P. (1999). Az 1970. és 1998. évi felső-tiszai árhullámok összehasonlítása, árvízi szimulációs vizsgálatok. (Comparative simulation analysis of the 1970 and 1008 floods of the Upper Tisza river) *Vízügyi Közlemények* 1999/3, 354–390. (in Hungarian).

Szöllősi-Nagy, A. (1989). *A mederbeli lefolyás real-time előrejelzése dinamikus strukturális-stochasztikus modellekkel.* (Real-time forecasting of channel runoff with the help of dynamic structural-stochastic models) VITUKI, Budapest, Hungary. (in Hungarian).

Todini, E. and Bossi, A. (1986). PAB (Parabolic and Backwater) an unconditionally stable flood routing scheme particularly suited for real time forecasting and control. *Journal of Hydraulic Research* **24**, 405–424.

7.4. Groundwater model in monitoring network design (Monitoring Design)

This example is based on a study carried out in the province of Utrecht, in the centre of The Netherlands. The province operates a groundwater head monitoring network and the objective of the study was to optimise the existing monitoring network by identifying the number and locations of observation wells required to achieve the desired accuracy. The objectives of the monitoring network were that it should provide information about the characteristics of the groundwater system, such as the spatial distribution of the time-average head and the average high groundwater head (AHG) defined as the 87.5% percentile. The accuracy was formulated in terms of standard deviation of the spatial interpolation of the AHG, i.e., the spatially interpolated AHG should represent the true AHG within a specified deviation.

A traditional approach to estimate the spatial distribution of the AHG would be interpolation by standard interpolation routines, such as Kriging. Kriging can provide both an interpolated AHG as well as and estimate of the accuracy of the interpolation (Kriging standard deviation), which is dependent on the data density and the geostatistical properties of the variable (variance and correlation lengths). As an alternative to traditional interpolation, a groundwater model was constructed for the area, which where then used to filter out the regional variation of the AHG. Kriging was thus used to interpolate the deviation between the observations and the model results, rather than AHG itself. Because the model accounts for the regional pattern in the groundwater head, the variability in the residuals between observations and the model is much less than the variation in the observed AHG, Figure 7.5. In the ideal case, with a perfect model, the residuals are small and spatially uncorrelated. Therefore, the better the model describes reality, the smaller the residuals, and consequently the smaller number of observation wells necessary to meet the required maximum Kriging standard deviation.

The basis for Kriging interpolation is the variogram, in particular the sill (indicating the total variability) and the range (a measure of the spatial correlation length). For the province of Utrecht we used 139 existing observation wells to calculate the variogram. In Figure 7.6a the variogram is given of the observed AHG without using the model and Figure 7.6b gives the variogram of the residuals. The sill in Figure 7.6a is over 22 m^2, whereas in the sill in Figure 7.6b is 0.32 m^2. This means that the model accounts for about 90% of the total spatial variability.

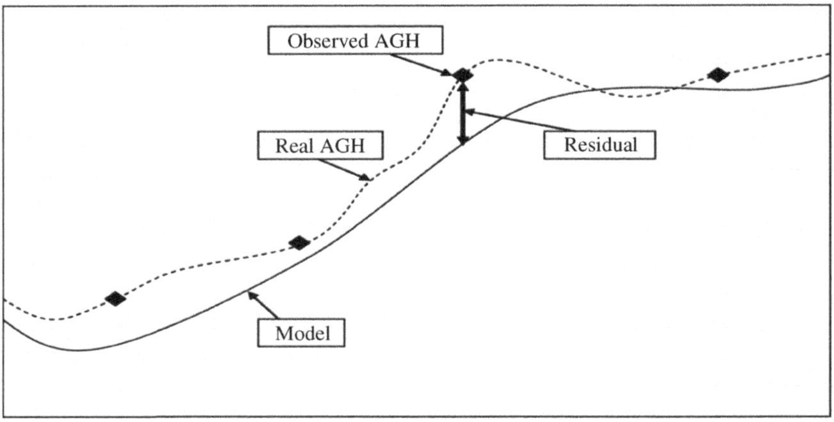

Figure 7.5 Illustration of the reduction in variance when the residual between observations and model results are considered instead of observations

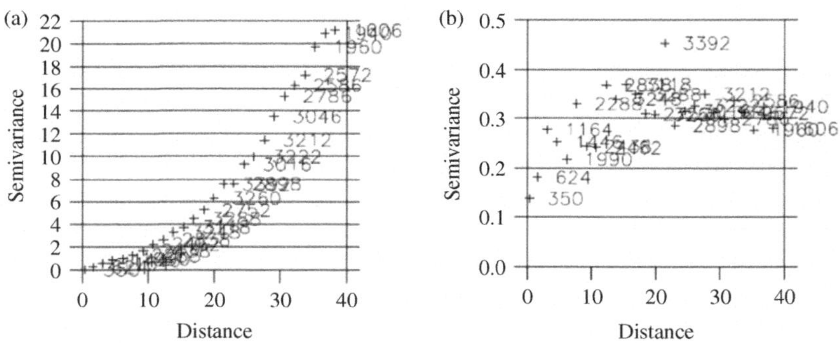

Figure 7.6 (a) Variogram of the AHG; (b) Variogram of residuals

The network lay-out was optimised by varying the number of observation wells and their location with the constraint that the Kriging standard deviation should be less than a given threshold. Using the model, with 40 monitoring wells, the Kriging standard deviation is less than 35 cm. If the model would not be used, the number of monitoring wells would be over 100 to reach the same Kriging standard deviation. Note that the model is used as a fixed reference. Once the model's ability to describe the regional pattern in the groundwater level has been mapped it can be used also in the interpolation for new monitoring periods, without running the model with new data.

8. APPENDIX – COMMON FLOW CHART

In 2006 and 2007 two workshops have been organized, where real life case studies from different areas around Europe have been elaborated to create flow charts on how to implement the WFD considering where to monitor and where to model and where these two disciplines should join forces. These flow charts have been combined to a common flow chart that we believe to be applicable in most WFD implementation challenges around Europe. The flow chart is described below.

A small summary of the six case studies are given here – more info can be found in Jørgensen *et al.* (2006) and Jørgensen *et al.* (2007).

Lithuania: Nevesis River Basin: Surface water quality problems; high nutrient loads from agricultural diffuse sources especially resulting in nitrate problems; besides several small villages without WWTP resulting in outlet of phosphorous.

Italy: Region of Campania: Groundwater quality problems, induces a severe problem to the drinking water supply in the areas, due to heavy agricultural activities (horticulture and live stocks). Besides outlets of untreated WW from small villages similarly contributes to poor groundwater quality.

Estonia: Pärnu River Basin: Potential threats towards wetland and the groundwater quality are caused by agricultural activities resulting in both diffuse and point sources – the latter due to improper handling and storage of manure. Besides there is some peat mining in the area resulting in local acidification of surface waters.

Denmark: Hørup well field: Groundwater abstraction for drinking water production – here the challenge is how to ensure sustainable abstraction without affecting nearby streams unacceptable. Besides there are some threats towards the groundwater quality due to diffuse pesticide pollution and point sources from contaminated sites in a nearby city.

Cyprus: Ezousas aquifer: Saline intrusion into the aquifer due to too heavy abstraction for irrigation. There is too little groundwater recharge to the aquifer as the precipitation in the area has gone down around 20% during the last two decades and also due to dam construction upstream a river running though the area.

Romania: Ialomita River Basin: Too high level of nutrients, especially nitrate, in both surface and groundwater from a range of different sources: Aerial deposition from neighbouring areas with agricultural activities; diffuse pollution from the local agriculture although the present pressure is low. Finally there is a special problem in terms of groundwater storage due to very high nitrate use in the past.

All these case studies have been elaborated by small groups in order to create flowcharts on how to implement the WFD in the given case, considering model support and joining this with monitoring activities.

Based on the constructed flowcharts we have proposed a common flowchart that we believe is applicable in most areas around Europe facing different challenges in WFD implementation.

The different steps in the flowchart (marked by numbers at the left side of the diagram) will be described below with reference to the six case studies above. Finally some general comments on further modelling support will be given and also some remarks on data accessibility. Terms related to the implementation will be in **bold**; activities relating to data collection *underlined and in italics* and modelling activities in ***bold and italics***.

8.1. Description of steps in flowchart

Step 1: **Problem identification**
When implementing the WFD in your River Basin the natural thing to start with is by formulating what the **problems** or challenges are –

> *Nutrients as in the Italian, Estonian, Lithuanian and Romanian cases, non-sustainable groundwater abstraction as in Cyprus, threats against and even from an important well field area as in the Danish case – or something completely different?*

This will be estimated on the basis of available data – which could be classified as *surveillance monitoring* data according to the terms used in the CIS GD on Monitoring.

Step 2: **Conceptual model**
The next thing to do is to establish a **conceptual understanding**. Here you combine all the available information about your system and make sure that all physical and chemical relationships are respected.

Here you must be sure about the physical parameters of your system – e.g.

- *The geometry of your streams (important in the Estonian and Lithuanian cases)*
- *The layers in your geology (important in the Cypriot, the Danish and the Romanian cases)*
- *The distribution of precipitation (important in all cases)*
- *Diffuse and point sources and how to distinguish these (as in the cases with nutrient problems)*

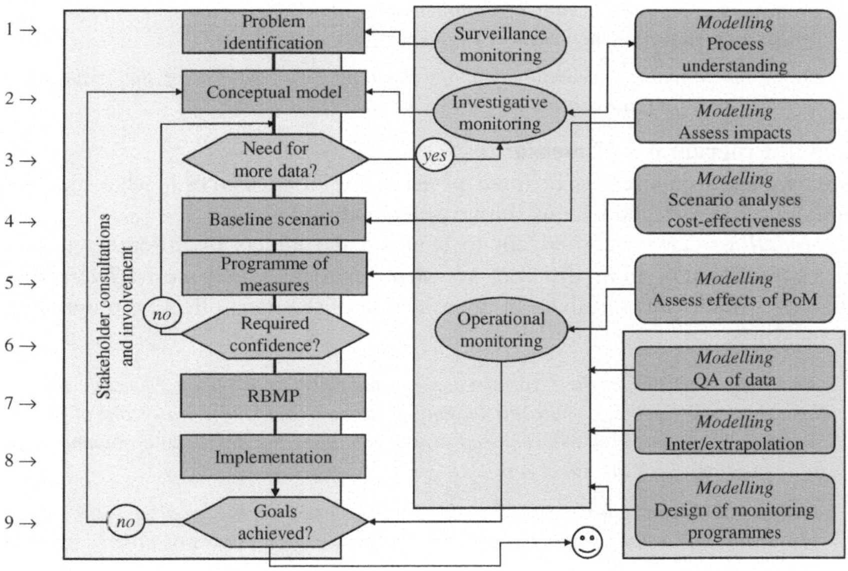

Here a ***model*** may help you to check the consistency of the ***processes*** in your conceptual understanding.

Step 3: **Need for more data?**
Now a decision needs to be taken: is there a need for *more data collection* – or are we confident with our conceptual model?

If we decide to carry out more monitoring this could be done as *investigative monitoring*, combining this with ***modelling*** and conceptualisation until we are satisfied.

As in the Romanian case, where there are several different sources of nitrate in the streams – the contribution from each source needs to be quantified by some dedicated monitoring campaign.

Step 4: **Baseline scenario**

Before the need for Programmes of Measures can be evaluated, the effect of measures already implemented or planned, as a result of other directives or initiatives must be evaluated. For such evaluation modelling is extremely useful, by using a ***model*** for running a **baseline scenario**, where measures implemented or planned are taken into account and the ***impacts*** hereof thus can be assessed. From the **baseline scenario** it can be evaluated whether additional measures are required.

As in the Danish Case – what will the effects be of already implemented programmes on remediation of contaminated sites? And what will happen if abstraction rates are increased or decreased?

Or in the Romanian cases – what are the long term impacts of the reduced intensity in agricultural practice – and what if this changes?

Step 5: **Programmes of measures**

Now we must choose one or more **programme of measures** – what must we initiate to achieve good status in our water body?

Modelling can be carried out to evaluate the effects of different possible measures. Results from different ***scenarios*** can help in a ***cost-effectiveness*** analysis, which can contribute in the selection of the most effective programme of measures.

An example of this is the Cypriot case – a measure on artificial recharge has just been implemented – will this be enough to keep the saline water out of the aquifer – or is more efforts required – and how does the decreasing amount of precipitation effect the situation?

And in the Italian case – would better farming practice in the area solve the problem – and what would be the socio-economic consequences of this??

Step 6: **Required confidence?**

At this stage a new decision needs to be taken – are we satisfied with the present level of **confidence**? What does our ***model*** tell us about the ***uncertainty*** of achieving our environmental goals?

If our confidence in the assessment is inadequate we need to go backwards in the process – to the conceptualisation or to the monitoring activities to see if we can improve the confidence.

As in the Danish case where it has been considered to move the well field to another location to reduce the impacts of abstraction on the streams in the area to an acceptable level. This is a very costly measure and before implementing this you need to evaluate whether you are confident in reaching the desired effects or you might need to collect more data and repeat some of the analysis and modelling tasks in these steps.

Step 7: **River Basin Management Plans (RBMP)**
When we are confident with our assessment we go straight forward to formulating our **RBMP**.

In this stage **stakeholder involvement** is of superior importance – but stakeholders actually need to be involved in all steps of the implementation process.

Step 8: **Implementation**
Now we can **implement** our RBMP. When doing this a number of measures are in operation and it is then of course very important to monitor whether these have the desired effects. This we call *operational monitoring* and this activity should be joined with ***modelling***, where the model is fed with new data as they are collected. In this way the model can at any time ***assess the effects of the implemented PoMs*** and reflect whether the measures are resulting in the desired effect or whether another strategy should be chosen.

Step 9: **Goals achieved**
The monitoring and modelling activities feed into the next decision whether we have **achieved our goals**.

If yes – then the right strategy was chosen. If we fail to meet the goals it means that the PoMs were wrong or insufficient, which indicate an inadequate understanding of the system, i.e., wrong or incomplete conceptual understanding. We thus have to go back and re-evaluate and improve our conceptual model, which may include additional investigative monitoring.

8.2. Additional Modelling Support

The above mentioned modelling tasks are linked to specific activities that have to be addressed in the implementation of the Water Framework Directive. Beyond these entries modelling can be supportive in basic tasks:

8.2.1. Quality assurance of data

In the Water Framework Directive and the associated guidances, quality assurance of data is restricted to quality assurance of sampling and laboratory procedures. Data may, however, be erroneous due to many other reasons. If however data are "brought alive" by using them in a model, inconsistencies and errors may be detected.

8.2.2. Interpolation/extrapolation

Monitoring data are discrete in time and space and often inter- or extrapolation techniques are required to obtain a continuous distribution of the variable. Standardised interpolation techniques are often applied, but these are limited by their generality. By the use of models, based on a sound conceptual understanding of the system, interpolation/extrapolation is based on physical/ chemical laws and the cause-effect relationship of the physical system. Model use in interpolation is therefore knowledge based and thus superior to standard techniques.

8.2.3. Optimisation of monitoring programmes

There are many different objectives to design and operate a monitoring system. The costs of the programme and the desired information level have to be balanced with the desired level of confidence or certainty. A model can be used to assess the uncertainty corresponding to various levels of measurement effort and in this way support an optimal design of monitoring programmes.

8.3. Data

Getting access to already collected data is not always an easy task, although both European and international conventions and agreements recommend free access to data collected for public funds.

As a final recommendation we will therefore like to urge water managers to make sure that all data collected in water management are stored in **public databases**; not only the data collected as a result of implementing the WFD but also other relevant data that can be of use and information for stakeholders, end-users, research tasks etc. This will facilitate the process of stakeholder involvement as well as encourage researchers to look into relevant challenges related to the WFD implementation.

8.3.1. References

Jørgensen, L.F., Brandt, G. and Vandenberghe, V. (2006). Joint use of monitoring and modelling when implementing the WFD. Case studies. Workshop report, Osnabrueck, Germany, April 2006. Harmoni-CA WP2/WP4, D4–5. Available at: www.harmoni-ca.info

Jørgensen, L.F., Brandt, G., Højberg, A.L., Refsgaard, J.C., Paukstys, B. and Zsuffa, I. (2007). Joint use of monitoring and modelling. Case studies for design of WFD implementation. Workshop report, Budapest, Hungary, June 2007. Harmoni-CA WP4, D4–6. Available at: www.harmoni-ca.info

Guidance Report II.4

Understanding the role of perception and valuation in the development and use of models for water management

Pieter Bots, Geoffrey Gooch, Brian S. McIntosh and Claudia Pahl-Wostl

Abstract

Understanding human decision making has become of key importance with the advent of more open regulatory frameworks such as the Water Framework Directive. In this document we address the human dimension of water management as human perceptions and valuations of the world around them, and as human decision making. Belief systems, human attitudes, and collective behaviours are presented as an integral part of both modelling and management objectives and strategies. The human dimension exerts itself through the filtering, selection and amplification choices. Perceptions and valuations imply that more than one legitimate and plausible interpretation of a problem situation may exist, in particular when the knowledge base is incomplete and uncertain. Given the recognised importance and potential value of using computer-based models to support many aspects of water management, it is the goal of this document to provide an overview of how the human dimension relates to and influences the development and subsequent use of models in water management. When addressing the model development process, we point out its non-linearity, the pivotal activity of model bounding and its sensitivity to the frames of modellers as well as other actors. While discussing the use of models, we focus on the task a model is to support, and the different organizational levels and modes of model use that need to be considered. This Guidance Report gives recommendations to three different types of actor involved in the development and use of models – the model commissioner, the model developer, and the model user – to raise their awareness of the human dimensions of model-supported water management.

Keywords

Human dimension of water quality management, mathematical modelling

 Integrated Assessment for Water Framework Directive Implementation Water Framework Directive Series Volume 2. Edited by Peter Vanrolleghem. ISBN: 9781843393269. Published by IWA Publishing, London, UK.

1. INTRODUCTION

Water resource management has a strong and successful engineering tradition based on controlling environmental problems with technical solutions. The management of risks has relied on the ability to predict extremes and control their impacts with technical measures (e.g., flood protection, design criteria for dikes, dams and reservoirs). Under this paradigm, belief systems, human attitudes and collective behaviours have been treated as external boundary conditions but not as an integral part of management objectives and strategies. However, the situation has started to change dramatically.

Complex demographic and economic processes are driving an increase in the number of households and urban growth across Europe (Ogden and Hall, 2004) with significant implications for water management. Tensions are apparent between the pressure to build households with reliable, high quality water, wastewater and drainage services and the recognized need to maintain sufficient water for environmental and ecological purposes. Areas experiencing the greatest economic growth may already have a negative supply-demand balance (Downward and Taylor, 2007) with implications for allocating water as a scarce resource. In addition, the incidence of flooding is on the increase as a consequence of pressure to build on land close to rivers combined with the spread of impermeable urban surfaces and changes to rainfall pattern resulting from climate change.

How should we manage our water resources? How should we seek to balance economic and demographic pressures, human water needs, environmental water needs within the context of a changing climate? Such problems manifest themselves through differences rooted fundamentally in human values and beliefs and have been termed 'wicked' (Rittel and Weber, 1973) for they admit no single formulation or solution. Indeed to even formulate such problems involves restricting the kinds of solutions that are to be considered e.g., viewing water resource problems as a function of demographics (too many people moving to areas where there is too little water) implies a macroeconomic solution. Should water management be framed as a set of primarily technological and infrastructural problems and solutions or should we be required to change how we use water or where and how we live? The answer will depend on the values and beliefs of the person being asked – there is an inescapable human dimension to the problems facing water resource management in Europe.

The importance of the human dimension of water management has been recognized and is reflected in for example, shifts towards demand management rather than the development of new supply side options insituations of water scarcity. Further, the importance of water governance and stakeholder and public participation in water management is now widely recognized. The European Water Framework Directive shows clear signs of a change in governance style with a requirement to develop and implement integrated management plans at river basin scale in a participatory fashion. But these new requirements meet a water management community largely unprepared to fulfill them.

The human dimensions of water management are also, crucially, reflected in the processes by which issues are identified as important and classed as problems to be solved. Processes of dialogue and debate between individuals result in changes to individual beliefs, the identification and formulation of problems and their potential solutions and the formation of collective (typically organizational) action. Scientific information is selected and used to support arguments during such processes and is subject to the influence of values and beliefs of those involved in dialogue, see for example the 'Processes for Organization Meanings' (POM) model of Checkland and Holwell (1999) which describes how information is acquired and used in organizational action. Belief and value based differences in the selection and use of information can lead to differences in the filtering and amplification of issues and the consequent formulation of problems and action oriented solutions.

Given the recognized importance and potential value of using computer-based models to support many aspects of water management from pure hydrologic modelling (e.g., Dorge and Windolf, 2003; Rekolainen *et al.*, 2003) through to integrated assessment of land-use change, water use and resource availability processes (Oxley *et al.*, 2004), it is the goal of this document to provide an overview of how the human dimension relates to and influences the development and subsequent use of models in water management.

By human dimension we understand:

HD I. Human perceptions and valuations of the world around them

Human actors hold subjective perceptions of the physical and social environment they live in. Subjective perceptions affect how and what kind of meaning is attributed to new information, e.g., about a change in the state of the environment. This implies that more than one legitimate and plausible interpretation of a problem situation may exist, in particular when the knowledge base is incomplete and uncertain. Actors differ as well in their preferences and values. While some attribute a high importance to the state of the environment, others give most weight to economic considerations.

HD II. Human decision making

Understanding human decision making has become of key importance with the advent of more open regulatory frameworks as the WFD. Decisions are affected by uncertainty in knowledge. Whereas water management has a strong tradition in dealing with quantifiable environmental uncertainties (e.g., variability in water flow) it has little experience in taking into account non-linear processes (e.g., breakdown of ecosystem function upon crossing thresholds) and uncertainties in human behaviour (e.g., unexpected change in behaviour upon introduction of new regulations leading to opposite effects). Much emphasis has been given to the development of decision support systems but has often been neglected processes during the phase preceding the final decision situation involved in determining the framing of the decision context and the options that are taken into consideration in any decision on water resource management issues.

Our target audience includes:

- practitioners involved in commissioning, using or developing models as a source of information or evidence for water management, and;
- researchers involved in the development of models for use by water management practitioners.

We have selected a broad and mixed target audience for a key aim with this document is to raise awareness of the human dimension aspects of model development and use. By reading the document it is our hope that the reader's critical appreciation of modelling and model use will be enhanced. It is our concern that models be seen as technologies for providing information and a means of improving water management outcomes, rather than as technological ends in themselves. As such this document is our attempt to embed models within and relate them to the wider set of human processes involved in developing and using them as a way of ensuring the emphasis of discussion clearly moves away from technology-push (models are the solution) to demand-pull (what can models do to improve water management?). We hope that both researchers and practitioners will better understand how models fit into the broader context of organizational and societal information acquisition, utilization and decision making processes involved in water management as a consequence.

We will start our document by clarifying what we mean by models and by looking at different ways in which models can be used for water management.

1.1. Models, modelling and the use of models in water management

We have set ourselves the goal of exploring the relationships between the human dimension, in terms of beliefs, values and decision-making, the development and use of models and the pursuit of improving water management. With regard to water management our aim has been motivated partly by the fact that models have been advocated as being suitable for supporting the process of implementation of the European Water Framework Directive. Specifically models have been identified as being relevant sources of support for WFD implementation through providing a way of:

- Setting operational targets for achieving the goal of 'Good Ecological Status'.
- Handling uncertainties and different problem framings when assessing the cost effectiveness of management options.
- Engaging stakeholders in processes of participation.
- Assessing the impacts of climate change uncertainties.
- Dealing with the complex cross-sector, multi-scale processes required to achieve integration in river basin management planning.

In this Guidance Report we will restrict ourselves to dealing with computer-based simulation models of any kind, used in any way to support an activity related to water management and in particular WFD implementation. By modelling we mean the process of developing computer-based simulation models and include in this process a range of activities from the social and conceptual (e.g., model specification and bounding) through formal (e.g., describing systems mathematically) to technical (e.g., implementing a model through developing a C++ program). Many of the issues and findings of this report will have some bearing on decision support systems (DSS) but this kind of technology will not be explicitly considered as it forms the subject matter of another volume of this series (Vanrolleghem, 2010).

By model-based water management we simply mean the use of computer-based simulation models to support activities related to water management. We will look at water management in broad terms as involving the management of hydrology, water quality, infrastructure and demand and examine more closely how models are used to support activities within these areas. In doing so we will cover a variety of uses of models from analysis to education and aim to speak to a range of readers from resource managers making operational decisions through water and wastewater service providers making strategic investment decisions to governmental policy makers determining national priorities. Importantly we will also speak to model developers, whether they are academics, consultants or located in utilities or governmental departments and agencies. We will cover both analytical and participatory uses of models in water management.

1.1.1. Types of models

There are many different ways of categorizing and classifying computer models based either upon their formal computational or mathematical basis (e.g., static vs. dynamic or deterministic vs. stochastic), their scale of representation (e.g., individual based vs. lumped) or their representational ontology (e.g., compartment flow vs. agent-based). There is no single universal way of categorizing models – the purpose of the classification dictates which dimensions should be used to differentiate between models. However Mitsch and Jorgensen (1989) provide a useful classification of types of model differentiated by a variety of attributes ranging from purpose through representational scale to computational basis.

Table 1.1 Summary of model types (Mitsch and Jorgensen, 1989)

Type of model	Characterisation
Research models	Used as a research tool
Management models	Used as a management tool
Deterministic models	Predicted values are computed exactly
Stochastic models	Predicted values depend on probability distribution
Dependent Compartment models	Variables defining the system are quantified by means of time-differential equations
Matrix models	Matrices are used in the mathematical formulation
Reductionistic models	As many relevant details are included as possible
Holistic models	General principles are involved
Static models	Variables defining the system do not depend on time
Dynamic models	Variables defining the system are a function of time (or perhaps of space)
Distributed models	Parameters are considered functions of time and space
Lumped models	Parameters are within certain prescribed spatial locations and times, considered as constants
Linear models	First-degree equations are used consecutively
Non-linear models	One or more equations are not first degree
Causal models	Inputs, states, and outputs are inter – related by use of causal relations
Black box models	Input disturbances affect only the output responses – no causality is required
Autonomous models	Derivatives do not explicitly depend on the independent variable (time)

More information on different models, types of models and ways of classifying models can be found by following the links below. Other resources can be found (there are many) by navigating to the links page of each site below.

- *http://www.euroharp.org/toolbox/* – EC funded Euroharp toolbox provides information and test results about a large number of tools/models aimed at quantification of nutrient losses from diffuse sources.
- *http://www.rbm-toolbox.net* – EC funded River Basin Manager's Toolbox providing information and tools for WFD implementation.
- *http://www.epa.gov/epahome/models.htm* – US EPA information sources site providing links to a range of models across specific environmental areas like water, air, transport, public health etc.
- *http://cfpub.epa.gov/crem/* – US EPA Council for Regulatory Environmental Modelling site providing links to libraries of models and tools for specific environmental areas like water quality and river basin management, and guidance on modelling standards.
- *http://water.usgs.gov/software/* – USGS water resources applications software library.
- *http://smig.usgs.gov/SMIG/model_archives.html* – USGS surface water quality and flow modelling interest group archive of models and modelling tools.

1.1.2. Use of models

There is yet little experience or accumulated knowledge regarding the influence of the human dimension on models and their use in the water management field. However, one can draw from experience in the field of Integrated Assessment which has developed considerable expertise in the integration of the human dimension into the development and use of models. The development of integrated assessment models (IAMs) has made considerable progress over the past years due to major breakthroughs in computation power (Parker *et al.*, 2002). However, despite remarkable technical advances the use of integrated models and decision support systems in policy and decision making processes has been argued to be unsatisfactory primarily as a result of not having taken different aspects of the human dimension into account adequately (McIntosh *et al.*, 2005). The emergent and fast growing research field of participatory modelling and group model building is one response to bridge this gap. The whole process of model development and implementation support processes of social learning where different perspectives are taken into account. Embedding modelling more closely into policy processes changes as well the modelling purposes.

Brugnach and Pahl-Wostl (2008) introduce a classification of different modelling purposes which provides a way of relating models and their use to human dimensions:

Prediction
Models may be used to predict dynamic characteristics of the system. In complex systems, prediction is typically not focused on the forecast of the development of a single variable in a specific system over time and space. Prediction refers rather to understanding overall system properties such as the effect of increasing diversity on the adaptive capacity of a system or the influence of network structure on the spread of innovation in a social. Hence such modelling exercises can generate general insights and support the development of guidelines for integrated system design (e.g., the role of centralized versus de-centralized control in resource management regimes).

Exploratory analysis
Models may be used to map the space of possible development trajectories of a complex system to find out more about unexpected behaviour or thresholds leading to abrupt change. In general it may not be possible to attach a number to the probability of certain development pathways, but simply providing evidence for its existence may be sufficient to trigger implications for resource management strategies. Scenarios should be developed in participatory settings to include the wealth of knowledge and creativity of human beings (Pahl-Wostl, in press; Van der Heijden, 1996).

Communication
It seems to lie in the nature of human thoughts to simplify complex systems. Models may serve the purpose to communicate insights on the complex dynamics to decision makers, stakeholder groups and the public at large. Hence models can serve an educational purpose to raise awareness and change deeply entrenched beliefs on the nature of system response. An example is given by difficulties to understand implications of positive feedback cycles or the danger to respond in a reactive way to observed changes when threshold effects may lead to abrupt system change (Carpenter *et al.*, 1999; Schlumpf *et al.*, 2001).

Learning
Communication in the sense of education may already be described as learning. However, what communication refers to here goes one step further. The model,

and in particular the whole model building process, supports a process of social learning and reflection in stakeholder groups. The model becomes part of the system it is supposed to represent (Pahl-Wostl, 2002). This implies an extreme shift in the role of the model and the role of those who guide the process of model building. Scientists adopt the role of facilitators participating in a process of co-production of knowledge rather than being "external observers" who reveal scientific truths (Vennix, 1996; Checkland, 1999; Sterman, 2000; Pahl-Wostl, in press).

1.2. Actors in model-based water management

So we know about the different kinds of models that are available and we know about the different kinds of uses they may be put to. This may appear sufficient but it is not. To sensitize the reader to what we mean by the 'human dimension' of model-based water management, and to the relevance of this dimension for those who are involved in either developing or using models for water management, we will sketch a possible variety of actor-types who may be involved. While these actor-types do not represent actual people, it is in being confronted with these stereotypes that the reader may recognize some of the issues that will be discussed during this report.

1.2.1. Archetypal actors in a modelling context

Academics

Seek to understand how the world works, and, as the world is very complex, tend to focus on one particular aspect of a problem. Love to find answers to questions that they have posed themselves. See models as descriptions of how the world ticks, and, as the world is very complex and often changing, always need money to collect new, more detailed data and/or build and test new, more detailed models.

Academics are raised in (or become adopted by) tribes (paradigms) who address similar questions using similar procedures. Rephrasing questions and questioning procedures is frowned upon (hence the word 'discipline'). The interaction between different tribes ranges from a display of respectful indifference (common between different physical sciences) to overt abuse (not uncommon between different social sciences).

Physical scientists prefer models that are based on what they see as well-confirmed theories. Ideally, models are constructed from elementary building blocks, such as the laws of Newtonian physics and should reveal some universal truth.

Economists have no problem with models based on what they see as reasonable assumptions (e.g., man is rational, differences average out, systems tend to a state of equilibrium). Prefer models that are sets of equations that can be solved analytically, or at least numerically in reasonable time. See mathematical elegance as an objective, rather than a constraint. In a choice between models and the 'outside world', they probably prefer models.

Social scientists with a taste for models will collect and explore large sets of data to test mere speculations. However, most social scientist prefer ten thousand lines of narrative description to a million lines of computer code. Understanding diversity is more important than generating universal knowledge, which is in any case, not possible according to them.

Computer scientists see models as sophisticated pieces of software. Like to construct 'meta-models', 'software development platforms', and 'integrated model architectures'. Consider model construction by other scientists as pottering because they lack in architecture. Try to impose standards that are 'state-of-the-art' and love to improve the state-of-the-art.

Experts
Seek to remain expert on their particular domain of expertise. Like to answers questions posed by others, provided that these questions pertain to their domain of expertise. Like to be involved in projects that allow them to learn more about their domain of expertise. Respect experts outside their own domain of expertise, but will contest the answers given by fellow experts when they deviate from their own. See models as mere tools: models will not answer questions; it's the experts who use them that do. Distrust any model unless they've constructed it themselves, or at least have worked with it extensively.

Systems thinkers
Found as academics and also employed as policy analysts. Seek to make sense of complex problems, ready to accept that there may not be a solution, in fact almost expecting there not to be a solution. Insist on reconsidering any given question in a wider context before even thinking of answering it. Insist on talking to a wide range of actors, especially when their client seems to be apprehensive about this. Will emphasize the need for an integrated model, once the question has been clarified. Know that all models are partial and often wrong, but will use them nonetheless when they are available. Are very fond of 'box and arrow' representations. Can, when sorely pressed, display a great capacity to reduce complexity, but will lose interest when there's less than five different feedback loops in the model.

Research funding agencies
Seek to allocate their budget in a way that gives the greatest chance for useful results, but will settle for visible results in more theoretical fields of research. Useful results usually taken to mean that someone somewhere outside a University seems at least slightly interested. Love to see research driven by questions that are societally relevant, but will settle for questions that are scientifically relevant. Love to see scientists from different disciplines and different countries work together, but will reject project proposals that allocate significant resources to problem definition and 'getting to know each other'. Have no questions themselves, other than those concerning time, budget, participating researchers, and how to measure output. See models as codified knowledge that can be transferred, and tend to favour projects that promise to set the standard for computer models in a given field.

Research project managers
When in the project proposal stage, seek a problem that on the one hand will be seen as societally relevant, and on the other hand will allow scientists from different disciplines to formulate questions that they find interesting. Once funded, have to justify their decisions in the eyes of funding agencies as well as scientists. See models as a means to integrate different disciplines, as products that are potentially useful for decision-makers, or (in the worst case) as deliverables that may impress the funding agency.

Decision-makers
May also be used to mean 'budget holders'. Seek to deal effectively with the problems in the area for which they are responsible. Are keenly aware of this responsibility and the need to be able to justify their decisions, often upon the basis of some form of cost reduction. Tend to pose 'how' and 'with what' questions, feeling that they already know the answers to 'why' and 'what' questions. See models as the embodiment of 'best available knowledge', but only when they trust those who constructed them. Rely on some expert in their staff (or an external consultant, or several external consultants) to use models to evaluate alternative options for strategic decisions. May use the output of models for routine decision-making.

Consultants
Seek to understand what are the questions that are really important for their client (usually a decision-maker). Prefer clients whose questions can be answered with available knowledge. See their own models as assets that have to generate revenue. Prefer simple, robust models, but will fight any suggestion that

they lack sophistication. Have a keen appreciation of the value of models and data sets that may be used again in the service of future clients.

Politicians
Seek to influence other politicians and decision-makers in ways that best serve the interests of their constituency. Also seek to be re-elected, as that is a precondition for serving the interest of their constituency. Know that the public likes clear messages, dislikes unkept promises, and has a short memory. Know that it is safer to speak of values than of specific goals, and that goals can be presented as means. Will never believe in a model, but appreciate models that support their views and play down on other models. Politicians who are also decision-makers will prefer policy options that are affordable and visible to the public.

Stakeholders
Organizations, that try to protect certain interests that they see as being threatened by oncoming changes (anthropogenic or other). Seek access to decision-makers and politicians to win them for their cause. Will appreciate any model that highlights the changes they fear. Will mistrust any model that seems at odds with their own perception.

The man and woman in the street
Are busy most of the time with having a life. Probably not even slightly concerned about water and its management unless water prices rise or drought restrictions on use come into force, in which case the way in which water is managed is clearly incompetent and ineffective. See politicians (often) and academics (sometimes) on television and may read about their views in the newspapers. See models as black boxes to be mistrusted unless they support their own perceptions. Can be fooled with statistics and graphs, but will immediately recognize their own place (or lack thereof) in models that display geographical maps or social groups. Tend to have a concept of quantities (notably money) that is inversely proportional to number of digits in which they are expressed. Tend to vote for politicians that propose simple solutions for complex problems.

1.2.2. Some typical interactions

Multi-disciplinary discourse:

- Agronomist: "Different cultures imply different seeding and fertilising practices that will affect ground water in different ways."
- Sociologist: "Indeed! Differences between cultures are bound to play a crucial role."

- Computer scientists: "Shouldn't we develop an agent-based model, then? To model cultural differences, I mean?"
- [*confused silence*]
- Hydrologist: "As we're interested in ground water quality only, why not just use our flow model, with its parameters set to reflect the results of your research?"
- Computer scientists: "The new hybrid model architecture we're working on will allow integration of both and flexibility to include more at a later date!"

In the middle of some policy process:

- Consultant: "If *that* is what you want, then you might also consider these other decision options."
- Decision maker: "What impact will these new options have on the other criteria?"
- Expert: "I can only guess, so I won't say."
- Academic: "We could answer that question, but we need a different model then."
- Systems thinker: "Preferably one that also takes into account the interaction with the other subsystems."
- Research project manager: "We have no time for that!"

Technology push:

- Systems thinker: "You will only be able to formulate policy once you've at least considered interactions between hydrology, land use change, soil chemistry, aquatic ecology, demographics, the price of wheat on the European market and beaver migration."
- Computer scientist: "We could integrate all of that in a modular, generic, graphical, extensible modelling environment with plug 'n play."
- Politician: "Will there be impressive graphics and maps that I can use to back up my existing opinion?"
- Decision maker: "Will the system be reliable and full training and technical support provided?"
- Research funding agency: "So, some decision makers and politicians helped shape the system? A great success!"

Demand pull:

- Decision maker: "I'd like you to tell me which management options are best and why."
- Academic: "Certainly, it will take 3 years and several hundred thousand euros to develop a new modelling approach."
- Decision maker: "I'd rather have simpler answers in 6 months."
- Research funding agency: "We'll give you money as long as you can develop, pilot, evaluate and commercially roll out your new model within 3 years."
- Research project manager: "Er, yes, we can do that ... (I think)."
- Politician: "Good to see that we are promoting practical, applied research."

1.3. The human dimension in model-based water management

After now having introduced a number of important issues, and taken a somewhat tongue in cheek look at different actors, we can see that the human dimension in model-based water management can be partly seen through the interactions of the actors involved. Our light hearted look at some stereotypical interactions is an attempt to expose the impact of differing aims and expectations between actors as a means of illustrating the influence of the human dimension in familiar terms.

However, human dimensions of modelling are much more than this, and we will now turn our attention to describing the nature and influence of the human dimension in model-based water management by distinguishing between the *process of developing* (designing and implementing) models and the *process of using* models. We will use this distinction between model development and model use to structure this Guidance Report.

1.3.1. Model development

The first process, development, is typically undertaken by researchers, specialist software or consultancy firms or perhaps by specialist technical staff within water management organizations. The human dimension here exerts itself through the filtering, selection and amplification choices which underlie the inclusion or exclusion of particular processes or variables at particular temporal or spatial resolutions in computer models. These choices form the bounding decisions which determine model scope and structure, and may be influenced by a range of factors like disciplinary background and expertise.

As pressures increase for water managers to utilize more complex models which integrate across sectors and scales and perhaps even include an explicit representation of human behaviour (e.g., water use) the need to appreciate and to actively manage human influences on model bounding increase. More complex IAMs typically represent sets of processes for which knowledge is uncertain and multiple model formulations may be equally plausible. Why should one bounding decision be preferred over another? How should the bounding process be structured and managed?

In addition to model bounding, the human dimension exerts itself through the choices which determine how a model is implemented in software terms. Implementation decisions can have implications for model formulation e.g., some implementation platforms or software packages may make certain types of model difficult to implement. Implementation decisions may also have implications for interoperability and compatibility with existing data systems or models. How should implementation be structured and managed?

Finally, during model development modellers may be required to represent human behaviour in simple (e.g., trend extrapolation) or more complicated (e.g., decision making processes) ways. Modelling human behaviour is an explicit attempt to formally represent and manipulate aspects of the human dimension in terms of values, beliefs, decision-making and action. What does it mean to represent human behaviour using a formal system like a computer model? What is involved and what are the limitations? Such questions are important to ensure that models are not used to generate spurious or meaningless information about human behaviour as parts of socio-natural water systems.

Section 2 of this Guidance Report describes the human dimension in the context of model development. It is aimed at:

- *practitioners* who want to or are involved in commissioning or using models and want to better understand, and better manage, the process of model development;
- *model developers* who want to gain a better appreciation of the intrinsically human, design aspects of model development with a view to improving their own design practices.

1.3.2. Model use

The second process, model use, is typically undertaken either by specialist technical personnel within water management organizations, by specialist consulting firms or occasionally by researchers commissioned to develop a model and use it to answer a set of questions. It is important to distinguish here between the human-computer-interaction (HCI) aspect of model use which covers the ergonomics of the interaction between a single individual and a single model on a computer, versus the task and organizational levels of model use which cover the way in which models are used as part of an individual's job and as part of the wider set of processes involved in using information as a basis for taking organizational action. We will primarily focus on the latter aspects of model use for the ergonomic aspects are well documented.

The human dimension in model use exerts itself through the way in which models are used as tools to generate and manipulate information as part of processes of dialogue and decision-making within organizations. Like model development, the information output of models is subject to processes of selection, filtering and amplification and it is important for model developers to appreciate these wider aspects of model use beyond simple notions of model output directly shaping water management action. The use of models, and perhaps more importantly the information they generate, is complex, sophisticated and not well understood.

Section 3 of this Guidance Report describes the human dimension in the context of model use. It is aimed at:

- *model developers* and *researchers* who wish to gain a better understanding of the ways in which information is acquired and used during processes of organizational decision-making with a view to providing improved information support tools (e.g., models).

We have now spent long enough explaining our aims and hopes for the guidance document. We hope you find our report stimulating and instructive. It is not our intention to provide a 'how to' manual for reflecting knowledge of the human dimension into model-based water management. The human dimension is too broad a topic to be able to do so adequately and concisely. Instead we have opted for a report model with a more modest aim: simply to promote critical practice in the development and use of models through improving appreciation of the human dimension. Enjoy!

2. HUMAN DIMENSION IN THE DEVELOPMENT OF MODELS

The human dimension of model-based water management manifests in two ways: through the influences that individuals and groups have on the development of models, and through the influences that individuals and groups have on the adoption and use of models and the interpretation of their output.

This section of the Guidance Report will explore the first aspect of the human dimension by first of all investigating how the frame-dependent perceptions of modellers can influence the design of models and how these influences might be managed, particularly where there are differences within a modelling team. This leads to a rejection of idealized descriptions of modelling processes in favour of a more realistic, 'human' view on how models are developed, and to some guidelines for managing differences in perceptions and frames. The second part of this section will address the issue of incorporating human needs, wants, values and behaviours into models, looking first at different ways of representing the human dimension in a model, and then at two approaches to obtain the information on *what* should be represented in a model: scenario modelling and participatory modelling.

The content of this section is targeted both at model users (clients) or commissioners (funders) (who may or may not be the same person/organization) and model developers (researchers, consultants or technical personnel within water management organizations). The aim in both cases is to raise awareness of how human dimensions can influence model development and to provide some advice on how to manage these influences.

2.1. Perception, framing and model development

2.1.1. Frames, beliefs and values

When thinking about the human dimension of modelling, the importance of perceptions can hardly be overestimated. The Thomas theorem "*If men define situations as real, they are real in their consequences*" (Thomas, 1966, p. 42) succinctly summarizes that humans act on the basis of how they perceive the world around them. Perceptions result from observations, of course, but it is important to realize that they are also constructed and reconstructed in communication with others (March, 1978). Actors communicate with each other about their observations and try in this way to make sense out of the complex and ambiguous world surrounding them (Weick, 1979).

Actors develop their perceptions through communication and sense-making, while 'filtering' and 'correcting' what they hear from others. In this context Rein and Schön (1991) speak of a complementary process of 'naming' and 'framing'. They describe 'framing' as the process of selecting, organizing and interpreting and subsequently giving meaning to a complex reality, in order thus to define the limits of knowing, analyzing, reasoning and acting. To facilitate 'framing', a second process takes place: giving a name to a situation which is experienced as problematic. This process, described as 'naming', is a contributory factor in the direction and intensity of the attention paid by actors: by giving it a name, a label, certain aspects of the situation are highlighted while others are glossed over and fall outside the field of vision.

Once an issue has been named and framed, new information congruent with this specific way of naming and framing is more likely to be attended to and subsequently recalled in comparison to new, incongruent information. When information relevant to an issue is unavailable, the actors involved will have a tendency to assume the missing information is consistent with the way the issue has been named and framed. When available information about an issue is ambiguous, actors will tend to distort the information to conform the way in which it has been named and framed.

A frame can be thought of as a system of beliefs, where some beliefs are more fundamental than others (Sabatier, 1988). Gray (1985), for example, stresses the importance of primitive beliefs in his theoretical model of the 'Ecological Attitude Domain'. The first of these primitive beliefs is that humankind is above and apart from nature, and that nature should be utilized by humans. The second is that "*progress and growth are natural, inevitable, and good*" (Gray, 1985, p. 47). Acceptance or rejection of these beliefs lies at the base of many environmental disputes, and primitive beliefs are "*thought to be some of our most deeply internalized and most determinative of behaviours*" (Ibid., 32). According to Gray, these primitive beliefs, together with other aspects of environmental disputes such as general environmental concern, with beliefs about the costs and benefits of individual or societal actions on the environment, and with beliefs about individual responsibility and rights, are 'primary beliefs'. These can be placed at the base of an environmental belief system, and lead in turn to derived beliefs concerning conservation, pollution, and population, and to general environmental attitudes.

The importance of primary beliefs in a person's belief hierarchy has also been stressed by Fishbein and Ajzen (1975), and, according to Heberlein and Black (1981, p. 248) a series of beliefs, which can be both cognitive and evaluative, can combine to create an attitude. Beliefs are also important in the formation of a value, which is an "*enduring belief that a specific mode of conduct or end-state of existence is personally or socially preferable*" (Rokeach, 1973, p. 5). Attitudes and values differ in that "*an attitude refers to an organization of several beliefs around a specific object or situation. A value, on the other hand, refers to a single belief of a very specific kind*" (Rokeach, 1973, p. 18).

When studying policymaking processes, Vickers (1965) also observed the influence of frames, beliefs and values, although he used a different terminology. Vickers was keen to understand the mutual and inseparable nature of the relationship between judgments that people make about reality (what is true, what is likely, what might happen) and the judgments they make about value (what is important, what matters, what to want), and how these combine to produce 'readiness' to emphasize and distinguish some aspects of a situation whilst de-emphasizing others.

The 'filtering' and 'correcting' role of frames in human communication can be explained by the tendency of individuals and groups to order beliefs and values in a consistent cognitive framework (Heberlein and Black, 1981), to reduce cognitive dissonance (Festinger, 1957). Where substantial cognitive inconsistencies exist (Rokeach, 1960), it is to be expected that pattern-making tendencies and the inclination to organize values and beliefs in personal and societal 'schema' will occur (Douglas, 1966; Anderson, 1980). Mary Douglas and other proponents of 'cultural theory' (e.g., Douglas and Wildavsky, 1982; Douglas, 1992; Thompson, Ellis and Wildavsky, 1990; Schwarz and Thompson, 1990; Dake, 1992) have proposed that such schemas can constitute coherent systems that include both societal and environmental values and beliefs.

As frames become more complex, they also become more resistant to change as the effort to accommodating new information increases with the number of beliefs that need to be adjusted to maintain frame consistency. This explains why actors can engage in 'dialogues of the deaf' (Van Eeten, 1999) where, although they are discussing the same reality, they cannot (or will not) make the effort to adjust their own frame to make sense of the other's perception, and hence will be unable to reach meaningful agreement on the issue at hand.

How then is this important for water managers and others involved in the development of models? Primarily, insights into the importance of frames, beliefs and values, and of the ways that information can be filtered, should lead us to look more critically at the variables that are included and considered important in specific models. Why have exactly those variables or indicators been considered as central? Which indicators and variables have been omitted, and why? How have the model developers' 'tribal' preferences influenced these choices? While it may be difficult for water managers to get to the bottom of these issues, it is important that they do ask questions about why and how certain factors have been preferred in the model development. By understanding that choices, conscious or unconscious, have been made, it is possible to judge the effects of these biases during the model use.

2.1.2. Re-humanizing our representation of the modelling process

As modelling practitioners we are all familiar with textbook and paper models of the modelling process. These typically take the form of some kind of flow diagram which presents a set of logically distinct activities connected in some form of sequence. Sometimes these activities are linear in sequence with few inter-connections and sometimes there are greater numbers of inter-connections and the flow between the activities may be non-linear, with some activities repeated. Sometimes the models are descriptive in nature (what happens in practice) and sometimes they are prescriptive (what should happen in practice). Figure 2.1 shows two examples of descriptive models.

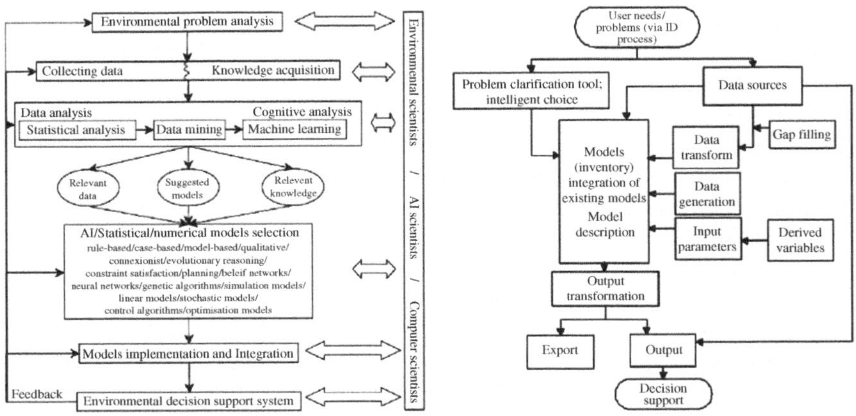

Figure 2.1 Schematic representations of modelling processes

Poch *et al.* (2005) (Figure 2, p. 860) present a descriptive model of how they developed two environmental decision support systems (EDSSs). The model involves 6 major activities ranging from 'environmental problem analysis' through 'models selection' to eventual implementation of the EDSS itself. The model involves feedback from the EDSS produced to each activity, presumably representing some form of learning by the model development team over time (for improvement of future versions of the EDSS, or entirely new EDSS). Inter-connections between each activity are not explicitly presented or discussed so the reader must assume the modelling process is linear and sequential. In addition, no discussion of the rationale behind the model is provided.

Argent and Houghton (2001) (Figure 1, p. 355) present a descriptive 'framework for water resources modelling' (a model of the modelling process) which starts from a set of user needs / problems and terminates in 'decision support'. There are 11 activities involved and some of run in parallel, but without any mutual influence, such as model selection and data acquisition. The authors recognize that the model is 'somewhat linear' and do not present an explicit discussion of the rationale behind the model.

The view of modelling presented is, at least implicitly, one of a process composed of distinct, sequential activities carried out relatively mechanistically rather than in a fundamentally human and social process. However, as modelling practitioners, we all know this not to be the case, particularly when developing models of complex, dynamic, multi-scale and sometimes socio-natural processes such as those involved in water resources, use and management. As practitioners we all know that modelling is difficult, messy and subject to a range of human influencing factors such as software technology availability and familiarity within development teams and the need for modellers to pursue employment incentive structures (e.g., publishing papers, developing their own models rather than using someone else's etc.). It is interesting to note that some water management users of models are aware of the influence that the HD can have on process and product. For example Caminiti (2004) recognizes that model developers can have different objectives from the model funder/user as a result of wider systemic issues e.g., University researchers having to contract out their services as a means of pursuing their own research agendas. Indeed, as recognition of the potential of models to act as tools to inform management action has grown, there has been a concern within the water and environmental modelling community to identify good modelling practice (Jakeman *et al.*, 2006) and to promote quality assurance (Refsgaard *et al.*, 2005). Figure 2.2 shows, as an example, the prescriptive model of the modelling process presented by Jakeman *et al.* (2006). It is composed of 10 iterative activities, each of which is described and discussed in the context of good practice. The authors recognize that modelling will involve trial and error, and repetition of steps, and that the boundaries between steps may in fact be blurred. Interestingly however, the overall process representation remains largely sequential in nature.

Such work has been motivated by recognition that the modelling process is not mechanistic or free from human contingencies and the risks of ignoring such issues may be significant and include model misuse through a lack of appreciation of various sometimes subjective choices involved in modelling (Jakeman *et al.*, 2006). However little research has been published on the modelling process itself – modelling practice and quality assurance work has been concerned with prescribing 'good' practice as a means of avoiding problems rather than exposing and understanding the underlying social process and wider systemic nature of modelling. What human dimensions influence the models we develop and the information and knowledge generated as a consequence? What implications do these human dimension influences have for how modelling processes might be managed?

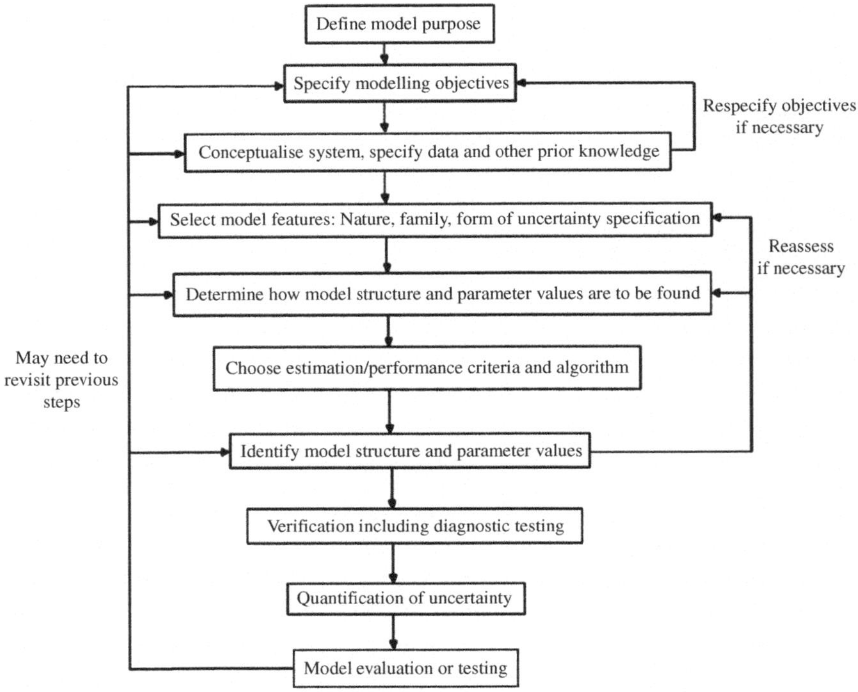

Figure 2.2 Relations between model building steps (Jakeman *et al.*, 2006, Fig. 1 on p. 607)

Some explicit research on the modelling process has been carried out within the domain of Operations Research (OR) by Willemain (1994). Through interviewing 12 'expert' modellers from a range of senior industrial and academic backgrounds Willemain asked about issues such as what makes a good modeller and what makes a good modelling process? He found that the qualities most frequently thought of as characterizing an effective modeller were related to mind-set (e.g., being creative, open-minded, accurate, sensitive to client needs) rather than technical expertise. The findings imply that effectiveness of a modelling process are at least partially dependent on a set of 'softer' human skills and competences. As modellers and educators of modellers we should be asking the question – do we have these skills and are we passing them on? Investment in the development of softer competences may improve the effectiveness of models and modelling.

One of the most mysterious of activities in modelling is model bounding (sometimes called problem definition or identification). The decisions made during model formulation are fundamental in that they set the bounds of a model (variables, relationships, scales) and therefore the questions that can be addressed through its use. For such a fundamental activity little is written about what is involved or how it should be carried out. It is typically presented as a black box activity despite its pivotal role.

Willemain (1995) investigated the first hour of the formulation of a mathematical model with the aim of understanding what modellers think about and when during the process of model formulation. Through plotting how modellers vary their attention from topic to topic over time (e.g., from topics concerned with context through to those concerned with structure, realization, assessment and implementation) he found that modellers varied their attention between topics constantly, particularly between 'structural' considerations and other topics. He found that formulation was like an entire modelling process in miniature and involved consideration of topics from all stages in model development. In addition, differences in style (sequence and proportion of time allocated to topics) were apparent between modellers. One of his conclusions is that the idea of distinct, sequential stages in modelling is dangerous if taken too literally.

In combination with his earlier work (Willemain, 1994) modelling can be seen as a craft rather than a mechanistic process – fundamentally individual and non-linear. This presents a challenge to typical models of the modelling process found in textbooks and papers. If modelling is not a sequential process, how does it work? What are the cognitive processes involved in modelling and how are they organized? What process differences exist between modelling undertaken by a single person vs. modelling undertaken by teams, perhaps within a collaborative/participatory context?

Research to date does not provide clear answers to these questions, but it is safe to say that the human dimension, in particular the frames of modellers, plays a crucial role in the modelling process, and that the process of model bounding is where we can expect the greatest effect of the human dimension, particularly with regards to complex environmental issues where uncertainty and equivocality are common.

2.1.3. Case study I – Examining influences on integrated assessment model development

So what do we know about the influence of the human dimension on model development, and in particular the model bounding process? In terms of published literature the answer is that we know very little but there are grounds to suspect that model development outcomes are contingent and heavily dependent on the composition of the modelling team. To counter the lack of published material we will instead examine a case study in model development. As part of the EC funded TiGrESS project (http://www.tigress.ac) research was undertaken to map out some of the human dimension influences on model design in the context of integrated assessment modelling (IAM). The findings have not been formally published yet but include a number of insights.

The field of IAM (Parker *et al.*, 2002) has emerged to provide model-based tools to help integrate "*knowledge from different disciplines with the goal to contribute to understanding and solving complex societal problems that arise from the interaction between humans and the environment*" (Gough *et al.*, 1998). Integrated assessment modellers have socio-natural systems as their focus with different elements of the IAM community focusing more on social or natural processes depending on background and substantive interests.

Important differences exist between the structure and rules of physical (e.g., soil chemical reactions), biological/ecological (e.g., community change), social (e.g., demographic) or socio-natural (e.g., food production, retail and consumption) systems. These differences are partly related to the degree to which a system can be described as mechanistic (however simple or complex) or evolutionary (Jeffrey and McIntosh, 2006). Physical systems are likely to be time invariant and mechanistic in that the rules underlying system dynamics will not change over time, although this does not mean they will be simple or easily predictable for particular cases. Biological/ecological systems include a degree of organizm learning and behavioural adaptation over shorter timescales and natural selection and evolution over longer timescales so there is no guarantee that a biological system studied now will follow the same rules in 40 years time – they are non-mechanistic (evolutionary). Social and socio-natural systems may change fundamentally so quickly that by the time an understanding has been generated about how they work they may have changed structure and rules and in fact no longer be the same 'system' – complex and potentially evolutionary *in extremis.*

This raises a number of issues relating to the human dimension of model development – given that socio-natural systems are always complex and potentially also evolutionary – how can adequate model bounding and design be assured? How do the disciplinary backgrounds and experiences of model development teams influence bounding decisions? Upon what basis should one model design (model ontology, variables and relationships) be preferred over another? Do any other human dimension influences affect model design? Do these influences matter?

It was clear from interviews held with 6 different IAM developers as part of the TiGrESS project that model design was subject to a range of influences, some technical, some non-technical and that the rationale behind taking particular design decisions varied qualitatively. Some non-technical (human dimension related) reasons for taking model design decisions included:

- To ensure professional acceptability of model [to fellow modellers].
- To maintain the support and interest of industrial partners.
- To delineate working responsibilities and manage work across partner institutions in the development team.
- To reduce the volume of work undertaken.
- To limit the number of novel aspects being tackled at once.
- To counter the risk of PhD projects not finishing on time.
- To move into an area where no work has been done before.
- To maintain a balance between contributing to the development of a new technique and contributing to substantive knowledge.
- To reflect the interests of the institution leading the development.
- To fulfill longer term personal and institutional ambitions.
- Because it allows searching solutions for an abstract class of questions.
- Because the proponent of the concept left the project.
- Because the expertise is available within the project.
- Because of difficulties in accessing relevant stakeholders.
- Because there is no clear additional benefit from alternative designs.

Model development and eventual design would seem therefore to be a function of (in no particular order) (i) balancing acceptability to other modellers with acceptability and understandability to (potential) clients, (ii) personal and institutional interests, (iii) available expertise, (iv) a need to include but to also manage and limit novelty, (v) managing workload and responsibilities within the development team and (vi) no clear objective benefits of choosing one design over another.

The significance and precise impact of these influences on the eventual design of any particular integrated assessment model is not clear but they do suggest strongly that modelling is an activity which could, perhaps even should, be viewed systemically as embedded in different personal and institutional contexts. Whether these influences should be or even could be managed out of the process is a point of debate but from the model commissioner's point of view, Caminiti (2004) was wise to acknowledge differences in aims that can exist between model developers and their clients and to suggest a process-based means of ensuring that these differences do not pose a risk to the delivery of a model that is usable and useful to the eventual user. Whether a tight process-based prescription for model development is feasible is a point of debate given the modelling-as-craft findings of Willemain. In the following section we will look into this more deeply.

2.1.4. Managing differences in perception and frame

For water resource management issues, there are good reasons for deliberately structuring modelling teams (depending on what is being modelled) to include a specific range of disciplinary backgrounds. Within such teams, differences in frames and appreciation will no doubt exist between *inter alia* different disciplines (e.g., chemists vs. hydrologists vs. anthropologists vs. economists). To avoid unintentional biases towards particular frames, it would seem a good strategy to deliberately engage in a period of brainstorming during model bounding to create a set of possible problem formulations before winnowing this set down using some criteria. The modelling process could be organized in such a way that the modellers iterate through divergent or diversity generating stages followed by convergent or diversity winnowing stages. However, when modellers with different frames have to decide on what aspects should be modelled and what questions the model should be capable of answering, they are likely to have different opinions. Defining criteria and decision rules for the convergent phases will remain difficult.

Problem formulation/model bounding can be seen as an example of 'appreciation' in the sense of the concept developed by Vickers (1965). Rather than prescribing a set of rules for bounding, Vickers was keen that the process of appreciation be made explicit through reflective documentation and a statement of the various judgements involved. Frames – by their nature implicit and not directly observable – will make that some issues are emphasized or included in models and others not. Making these frames explicit can reveal the rationale for the particular appreciation of the world which any model constitutes. This will allow not only the modellers but also other actors (notably project managers, funding agencies, politicians and other stakeholders) to have a meaningful discussion about the problem, the model and its use.

There exist no straightforward recipes for designing and implementing processes involving divergence-convergence and deliberate 'frame reflection', but some guiding principles can be found in the literature. Having reflected on experiences in several inter-disciplinary modelling projects, Nicolson *et al.* (2002) propose ten 'rules of thumb' that may help to meet the challenge of constructing a model that integrates different disciplinary perspectives:

(1) Select team members who can and will "*creatively explore the linkages and interfaces between their own discipline and other fields of inquiry in which they may themselves have no special expertise*", who have "*the ability to simplify what is known and, when necessary, guess at the unknown*" (see also rule 8), and "*whose interest in a problem outweighs their concern for career advancement*" (because interdisciplinary research usually does not lead to publications in high-ranking disciplinary journals).

(2) Develop a shared view of the research problem in an early stage (using rapid prototyping, see rule 3) through an explicit *process of negotiation* in which all parties express their perspectives and learn about each other's priorities.

(3) Use rapid prototyping for *all* modelling efforts. Instead of drawing up specifications of the final model, jointly develop a prototype model for the explicit purpose of clarifying the objectives of the study. The prototype model and its output reveals the big picture, gives an idea of the system dynamics as a whole, tests initial hypotheses, places individual contributions in perspective, and allows assessment of the relative importance of different system components. Refine both problem and model further through successive prototypes.

(4) Allow the project's focus to evolve by not allocating all funds up front. Though at odds with most institutional funding policies, this rule must be followed to provide the necessary conditions for the preceding two rules.

(5) Ban all models or model components that are inscrutable, as such 'black box' models impede meaningful exchanges between participants. By contrast, encourage the use of basic 'box and arrow' diagrams and spreadsheet models that "*allow scientists from different disciplinary backgrounds to understand and engage with the key relationships of the model.*"

(6) Instead of concentrating on one all-purpose synthesis model, invest in a suite of models, each with a well-defined objective. This will help participants to focus on the interfaces between different disciplines. Note that this rule does not mean that an integrated system model at the conceptual level (the 'box and arrow' diagrams).

(7) Maintain a healthy balance between the well-understood and the poorly understood components of the system. Note that deciding where 'educated guesses' are acceptable (taking away the opportunity for

in-depth analysis that some researcher might want to do) and which sub-models are too detailed (disqualifying the 'pet model' of some researcher) requires strong project leadership by someone who can see the big picture and is trusted and respected by the other team members.

(8) Sensitivity analysis (testing not only different parameter values but also the assumptions and the effect of alternative educated guesses at the underlying processes) is vital at all stages of the modelling effort. Nicolson *et al.* (2002) found that "*the ability to work as a group to set up a model simulation, and then view the results within a minute or two, was principally responsible for most advances in developing model relationships that crossed disciplinary boundaries.*"

(9) Work hard at communication and budget for face-to-face meetings. Communication between all parties involved is a necessary condition for most of the preceding rules. Although modern communication technologies can greatly facilitate exchanges, face-to-face meetings are invaluable for detecting and resolving misunderstandings and false consensus due to ambiguities.

(10) Approach the project with humility. Participants must be open for knowledge contributed by non-scientists, for being challenged by team members outside one's own area, and for learning from such transactions.

These rules of thumb first of all emphasize the need for *negotiation* between modellers. People's frames are resistant to change, so when actors merely exchange views this will have little effect on their perceptions of what should go into the model, and how. Putting effort into making model bounding decisions explicit, opening up 'black boxes' and focusing on decisions that lie on the interfaces between disciplines will help modellers to discover and make explicit where they have 'common ground' (Clark and Schaefer, 1989) and where they need to do grounding. Still, rule #1 and rule #10 make clear once more that the capability and willingness of actors to adapt their frames are crucial conditions for success.

Even when these conditions are met and the time and resources to negotiate are available, the intellectual challenge of bridging conceptual differences between models from different scientific disciplines remains. Although they do not resolve the problem of inconsistencies between models that are based on different disciplinary paradigms, multi-model software platforms (see e.g., Quesnel *et al.*, 2005) may facilitate the integration of different model types (e.g., the differential equation based models and agent-based models that will be discussed in Section 2.2.1). This brings to the fore that besides discipline-related frames, other frames may lead to conflicting opinions on modelling decisions. When, for example, different research groups each have invested much effort in their particular modelling software platform and have become experts in its use, they will be reluctant to switch.

Complementary to Nicolson *et al.* (2002) who focus on the challenge of interdisciplinarity from a research team leader's point of view, Caminiti (2004) takes the perspective of the commissioner of a model development project. As we argued in the previous section, model development is a intellectually demanding, often unpredictable iterative process. It is also a costly endeavour, and the required effort, time, and hence budget are easily underestimated. How does a water resource manager in this role obtain objective, independent information to decide what modelling approach to follow?

Caminiti (2004) emphasizes the risk of the model commissioner lacking the knowledge to judge the merits of different modelling approaches proposed by competing experts. To deal with this risk, she advises to "*seek technical assistance to develop a brief that specifies the water resource manager's objectives, the model's use, scope and boundaries, available data sources, compatibility requirements with other models, documentation of model inputs, format of model outputs, assessment of model uncertainties, ownership of the model and supporting documentation, description of scenarios to be tested, the level of stakeholder participation, communication of the model development and outputs, reporting arrangements, budget constraints, and a clear end point.*"

Another risk is that the water resource manager ends up serving the model rather than the model serving the water resource manager. Like our academic stereotype in Section 1.2, modellers tend to focus on model validity, rather than the water resource manager's needs. For them, the model may be a goal in itself, and because modelling often involves discovery and learning, modellers will ask for additional resources as the project unfolds. This risk of 'entrapment' is inherent to model development as an iterative process, but it can be managed by informed project specification and tender development, and by adopting a staged approach with (sub)model prototypes as intermediary products. To ensure continuity, the model commissioner should also protect the interests the funding agency and other key stakeholders and involve them in key decisions, for example by inviting them onto the project steering committee.

Then there will inevitably be uncertainties in the model and its output due to bounding decisions, quality and quantity of data, technical constraints, and/or the (implicit) assumptions in the management scenarios. Both model commissioner and stakeholders (especially when not intensively involved in the process) may fail to take these into account, and model output may be misinterpreted. The best strategy to mitigate this risk is to acknowledge that of computer models of complex phenomena are sets of assumptions that reflect the 'current understanding' of the system under study. Making modellers document and present these assumptions throughout the model development process mitigates the risk of both model commissioner and model developer coming to 'believe in the model', disregarding uncertainties and limitations. Even when data and knowledge are available, it is advisable to require such documentation so as not to let the model become a 'black box'. Ideally, the documentation is presented in a didactic form that allows non-experts (i.e., all actors besides the model developers themselves), when confronted with the model, to be able to understand how it works and what it can and cannot do.

Finally, Caminiti advises model commissioners to manage the expectations of stakeholders from the start, emphasizing the limitations of scope and use. Whenever possible, the perception of the model as a supporting tool for decision making should be enforced, and the perception that the model will provide solutions to problems should be contested.

It may be encouraging to see that many of the recommendations made by Nicolson *et al.* (2002) are quite compatible with those made by Caminiti (2004): avoid that models become 'black boxes', manage uncertainties explicitly, and develop complete resource management scenarios to test the model. The strategy of developing a suite of models, rather than one monolithic model, in order to be able to manage the interface between different disciplines will chunk up the project and help to manage the project financially.

By contrast, rules #3 and #4 of Nicolson *et al.*, are at odds with Caminiti's recommendation to develop a brief that completely specifies the model in an early stage of the model development process. Even with independent expert support, it will be difficult for the model commissioner to draw up complete specifications up front. Developing these specifications through rapid prototyping would seem to offer a compromise, balancing the need for project scope and budget control with the need for learning and discovery. 'Value Analysis' as proposed by Keen (1981) and definition of 'rules of the game' for deciding on whether or not to follow up on new ideas (Bots and De Bruijn, 2002) can help to structure and manage this prototyping process.

Having discussed how to deal with differences in perceptions and frames amongst model developers, and between modellers and model commissioners, we should not forget that on the side of the model users, political interests may lead to conflicting views on how the model should be bounded (cf. the stereotypes we gave of decision-makers, politicians and stakeholders in

Section 1.2). The following case study will include this particular human dimension while making a modest attempt towards generalization.

2.1.5. *Case study II – How frames influence model bounding decisions*

Can we somehow anticipate on the frames of the actors involved in a modelling process? And can we say anything in general of how particular frames will influence model bounding decisions? This second case study suggests that on a global level we can. It also confirms the need for explicit negotiations between actors when crucial model bounding decisions are being made.

Karstens *et al.* (2007) investigated how the choice of spatial and temporal scales was made for integrated modelling projects. They analyzed the Long Term Vision Study (LTV) of the Scheldt Estuary. The LTV mainly focused on the problem analysis and the design of four interim development sketches for achieving the Long Term Vision for the Scheldt Estuary in 2030. The impacts of the activities associated with these interim development sketches – notably deepening the main channels to improve accessibility of the port of Antwerp – were assessed.

The Scheldt river flows through three countries (France, Belgium, The Netherlands), so model bounding decisions concerning spatial scale would strongly affect the outcomes of the assessment. The assessment would also be very sensitive to the choice of time scale: economic impact models typically have a time horizon of several years, while ecological and morphological impact models typically have a time horizon of several decades.

The study revealed that conflicting opinions on which scales to choose were not only due to differences in scientific disciplines. To understand scale choice dilemmas in integrated assessment studies, Karstens *et al.*, identified three more frames in addition to the *scientific* frame: *political* (stakeholders having particular interests that might be affected by the model and its use), *managerial* (the commissioner of the study wanting to obtain useful results on time and within budget) and *analysis design* (the analysis team leader responsible for an integrated study):

- The 'scientific' frame of researchers/modellers will lead them to prefer modelling choices that will increase the scientific validity (to the eyes of their disciplinary peers) and interestingness of the study, for example because it involves investigating new phenomena or building more sophisticated models of conventional phenomena.

- The ‘political’ frame of policy makers and other stakeholder is dominated by the need to protect their own interests and preferably gain some benefit for their constituency and/or realize their personal political ambitions. This frame will make policy makers prefer modelling choices that will increase the possibilities for consensus building and issue trade-off, while it will make stakeholders prefer modelling choices that will give ‘their’ stakes prominence (or hides issues they consider to be too sensitive).
- The ‘managerial’ frame of the commissioner of the modelling project to an analysis team will favour modelling choices that will keep the time and resources needed for the study within certain limits, minimize risk, and maximize the likelihood of concrete results and actionable recommendations.
- The ‘analysis designer’ frame of the leader(s) of the modelling team will focus in particular on decisions that affect model bounding, trying to achieve systemic integration of the relevant aspects while maintaining a balance between the other three perspectives: meet the client’s wishes within the constraints of time and budget, be sensitive to political issues, but also to produce a study of good quality by scientific standards. This frame will lead them to favour those choices that limit the complexity of the model, enhance its capacity to assess a broad range of policy options, and offend neither stakeholders nor scientists.

The study by Karstens *et al.* (2007) shows that frame analysis can help to identify and clarify dilemmas that need to be addressed when making scale choices. Their approach seems applicable to other model bounding decisions as well, and may support the negotiation process referred to in Section 2.1.3, notably rule #2 of Nicolson *et al.* (2002). The four frames are indeed generic – note how the 15 reasons for taking model design decisions mentioned in Case study I can all be mapped onto one of the four frames – and provide a basic structure for explicit frame reflection during a modelling process.

2.2. Incorporating human needs, wants, values and behaviours into models

2.2.1. Representing the human dimension in a model

Although the technical aspects of model construction are beyond the scope of this document, we cannot totally disregard the human dimension of what is ‘inside’ a model, that is to say: the way human behaviour is represented in computer-based simulation models. Having some basic knowledge of these representations and the way they reduce (often drastically) the complexity of social systems should help the reader understand why models to support water resource management can be very controversial.

In Section 1.1, we argued that the variety of types of models defied the definition of a universal typology, and that one should choose categories that suit the purpose of the classification. For the purpose of illustrating different representations of human behaviour, we distinguish here between equation-based models and agent-based models. Equation-based models represent the relations between measurable system characteristics that reflect the aggregated effect of many individual behaviours whereas agent-based models represent these behaviours at the level of the individual (Parunak *et al.*, 1998). Thus, equation-based models represent the behaviour of large numbers of people in much the same way Avogadro's gas law $P*V = n*R*T$ represents the behaviour of large numbers of gas molecules: macro-properties such as density and flow of populations, and their pressure on the surrounding system can be approximated quite well using equations. The values of the parameters of these equations are determined by empirical observation and aggregation of the data using statistical methods.

To give a simple example: when looking at the issue of ensuring adequate provision of drinking water, the most prominent human dimension in this issue – the demand for drinking water – can be represented by averaging the water consumption behaviour of a great many people into a single value C (average drinking water consumption per year), and define the demand as $D = C*P$ with P being the population of the area under consideration. The values for C and P would depend on local characteristics (for Belgium, for example, $C = 125$ litre per person per day and $P = 10$ million persons, whereas for Italy $C = 210$ and $P = 58$). The variables C and P can be thought of as ultra-compact models: the value of C represents 'the average person' – the average of millions of people each having their own habits of drinking, showering, flushing toilets, sprinkling their lawns, and so on.

Like models of strictly physical phenomena, equation-based models of social phenomena will include equations that reflect theoretical assumptions. Some market models, for example, assume that supply and demand (two forms of human behaviour!) for some good will develop an equilibrium, and this assumption plus sufficient empirical data to estimate elasticity parameters allow to calculate at which price level the market will stabilize.

When differences between groups (defined socially, geographically or otherwise) are relevant for model use, sets of equations with different parameters per group can be constructed. Equation-based models become inadequate when

- the system population is heterogeneous in terms of attributes, actions, and decision rules,
- the topology of the interactions among individuals is heterogeneous and likely to produce clustering and other network effects, and/or
- the behaviour of individuals in the system is involves learning and adaptation (Bonabeau, 2002).

Agent-based models allow what 'averaging' equation-based models do not: to represent human behaviour from the individual's perspective, making explicit their particular way of deliberating and acting. Individuals are represented as autonomous decision-making entities ('agents'). At each time step during model execution, each agent individually assesses its situation and decides, usually on the basis of a set of goal-directed rules, which action (for example, moving about, producing, consuming, dealing with other agents) to perform. Agents may have some form of memory and additional rules that, based on past events, modify their decision-making rules, allowing adaptation that leads to new behaviour patterns.

To give a simple example: an individual's drinking water consumption depends not only on his/her basic needs, income, and having a garden or water-saving appliances (all these factors could also be included in equations), but also on social network characteristics, such as the use of water by other people in their social circle. To model this social influence, agents can be programmed to adopt behaviour rules of their 'neighbours', and this is likely to reveal dynamic patterns of water consumption that would go unobserved when using an equation-based model.

Agent-based models may vary in the way the agents' behavioural mechanisms are represented (Nickles *et al.*, 2004). The simplest agents link perception directly to action by a set of stimulus-response rules. More sophisticated agents have some evaluation function that allows them to compare alternative actions and choose the one that maximizes utility. Even more sophisticated agents deliberate on the basis of beliefs, desires and intentions (BDI). This representation allows the modeller to define agents with incomplete and even conflicting beliefs, and makes it possible to trace an agent's motivation for taking a particular action. Agents become socially aware when, in their deliberation, they also consider the beliefs, desires and intentions of other agents. The most sophisticated agents have the capability to learn (through feedback or through communication with other agents) and to coordinate their actions with those of other agents.

There is a broad range of projects which give examples on how agent-based-modelling can support water management. The box below gives some links to only a few examples.

FIRMA – Freshwater Integrated Resource Management with Agents http://firma.cfpm.org/ This project explored new approaches to improve water resource planning by developing and applying agent-based modelling to integrate physical, hydrological, social and economic aspects of water resource management. Specific emphasis was given to stakeholder participation and participatory model building and scenario development.

CAVES http://cfpm.org/caves/CAVESWiki/Summary?from = Main.Home-Main.HomePage
Complexity, Agents, Volatility, Evidence and Scale. The purpose of the CAVES project (2005–2008) was to couple policy concerns for complex human-environmental systems with linked physical, biological and social models based soundly in complexity science. They produced a constructive demonstration of modelling procedures for the formation of social policy in conditions of uncertainty due to complexity. To do so, a set of case studies of land use change driven by a variety of physical, biological and sociopolitical phenomena was developed. One case study is placed in the Oder River Valley in Poland, which has undergone considerable changes in land use, water regime and social structure over that period. The case studies provide extensive qualitative and quantitative evidence that constrain clusters of agent based models implemented for each case study. The models in each of these clusters include rich cognitive specifications of agents representing different stakeholders and models with more agents but simpler representations of cognition. Some models aggregate stakeholder groups. Others abstract from the detail of the case studies in order to provide simpler models that may yield some closed form solutions. Several prototype models are public available http://cfpm.org/caves/CAVESWiki/Models

FEARLUS http://www.macaulay.ac.uk/fearlus/index.html
http://www.macaulay.ac.uk/economics/research/Water.html
FEARLUS (1998–2003) was a response to the difficulties of understanding land use change, and hence of forseeing the likely consequences of possible

changes in the complex network of processes that produce it. In the project agent-based simulation modelling was adopted as the central component, using the Swarm (www.swarm.org) system developed at the Santa Fe Institute. The initial case study focused on the implementation of the WFD. It aimed to help identify the circumstances in which various intervention strategies (such as fines or incentives) are effective in preventing over-exploitation. Concurrent development of distributed multi-pollutant transport and impacts models was applied to provide the necessary process-level model input.

The following examples illustrate the complementarity of equation-based and agent-based models when modelling human behaviour in the context of water resource management.

Trends in well-defined indicators

When time series data on some human behaviour such as water use are available, a quick and simple method to derive a trend line is to perform a regression analysis to determine the parameters for the equation D(t) = *t +. However, a simple non-causal relationship ignores the real drivers affecting water use, and this type of model usually fails to properly predict human behaviour even when good quality data is available. A more disaggregated approach to demand forecasting using multiple linear regression model might be suitable (EC, 2003, pp. 149–150; Williamson *et al.*, 2002). For this kind of equation-based model, the key drivers of demand (e.g., temperature, rainfall, or garden surface) need to be identified and data on these drivers must be available. Interestingly, recent study by Gato *et al.* (2007) confirmed the researchers' hunch that water demand is for a large part fixed by basic need that are independent of external factors such as temperature and rainfall, with thresholds that are affected by weekends and weekdays. In this and similar studies, an R^2-value (the proportion of the variance explained by the independent variables) of 70% or more is considered to be very good. It is good to realize that the model equation(s) represent human behaviour not only as 'average' behaviour, but also do not represent the behaviour of 30% or more of the population under study.

Spatial development over time

Cellular automata models are somewhere in the middle of the equation-based – agent-based spectrum. They are grid-based, homogeneous agent-based models, meaning that they consist of thousands of square or hexagonal grid cells with a fixed location in a large array that represents a geographical area, and that each cell has the same set of rules – typically some relatively simple arithmetical equations. During model execution, these rules compute for subsequent time steps the state variables for a cell, such as population, land cover, water resource availability, etc. The equations usually define the cell state for time step t in terms of the state of the cell itself as well as the state of the neighbouring cells at time $t-1$ (Parker *et al.*, 2001). The properties and dynamics represented by the model can be physical (e.g., soil type and the flow of ground water and surface water), biological (e.g., the type and the spread of vegetation types), as well as social (e.g., migration and urbanisation), and their interaction between (e.g., loss of biodiversity due to desiccation, or erosion as a function of land use and precipitation) can all be represented in the same form. This facilitates the construction of integrated models, as well as the multi-disciplinary discourse on their validity and usability.

Stakeholder perceptions and positions

Water resource management involves decision making at different administrative levels, and the decisions involved affect the interests of a variety of stakeholders. Moreover, the cooperation of these stakeholders is often needed for effective implementation of the decisions that are eventually taken. To take into account these interests and to anticipate on stakeholder behaviour, the perceptions and positions of these stakeholders can be represented to some extent in computer models. Data obtained using interviews or other elicitation methods afford equation-based representations of the expected influence of actions on decision criteria (Bots, 2007a,b; Giordano *et al.*, 2007; Hermans, 2005) that help foresee conflicts and strategic behaviour (see Case study V in Section 3 for an example). The same data can also be used to define goals and decision rules for agents (Hare and Pahl-Wostl, 2002).

Adaptive water policy making

When sufficient insight in actor behaviour is available, the mutual dependence between the makers of water resource policies (the 'leaders') and those who have to abide by these policies (the 'followers') can be represented using 'two-level game' models (Putnam, 1988; Richards and Singh, 1997). These models simulate rounds of a certain time period. At the beginning of a year, leader agents define policies, for example by allocating water quota to particular uses, and by setting different tax rates for different industries. The various types of follower agents (industries, households, etc.) then act in order to maximize their goal attainment under this given policy. At the end of the period, the leaders observe the outcomes (the state of the water resource, economic productivity, etc.) and will, when these outcomes do not meet their policy objectives, adapt their policy for the next year. Two-level game models are particularly useful for exploratory analysis. Moreover, they can be effective tools for communication and learning, especially when real people can 'play' certain agents (leaders and/or followers), thus effectively incorporating a genuine 'human dimension' in the model (Barreteau, 2001; Guyot and Honiden, 2006).

These brief examples – presented here in order of increasing knowledge on the side of the modeller – not only show the variety of human behaviour that can be 'put into a model', but also underline once more the approximative nature of models in general, the number and variety of decisions that modellers have to make when developing a model, and the strong influence of these decisions on a model's usability.

2.2.2. Scenario modelling: how perceptions influence scenario modelling and use

A type of modelling which has received a considerable amount of interest during recent years is scenario-building. Scenarios have been used by industries such as oil companies, politicians, the military, and many others. As a form of modelling for water managers they represent a potential tool of significance. However, scenarios are as susceptible to the influences of human dimensions of modelling as other models. It is important to remember that models and scenarios are based upon human perceptions of the future, as well as human activities, and that these are incorporated into scenario modelling. The choice of factors that are included in scenarios naturally influences their outcomes. During recent years the use of scenarios for environmental policy-making has attracted considerable attention from both the scientific community and policy-makers (see, for example, (Greeuw *et al.*, 2000). Scenarios are not precise predictions of the future (Porter, 1985), they should be seen as simulation tools, as a technique similar to, but different from, models (Jouvenel, 2000). Scenarios attempt to construct general representations of possible alternative futures and they can be used to present the results of relatively complicated scientific modelling exercises to stakeholders and the public in a manner that they are able to understand. Scenarios, through their simplification of the information produced in modelling, can make this process easier. Basically scenarios can be qualitative or quantitative, explorative or normative. Often 'storylines' or 'scripts', are first developed (Gooch and Stålnacke, 2006); these can be narrative texts that act as representations of possible futures. These do not contain detailed numerical information, and can be written in an easily understandable manner for policy-makers, stakeholders, and the public. Qualitative/explorative scenarios can then be used as input into computer-generated models that can best be described as quantitative/explorative in the first stage, and quantitative/normative in a second stage.

Finally, the results of the quantitative models can be used to modify the qualitative scenarios, and to develop qualitative/normative scenarios that provided grounds for policy recommendations.

Human perceptions influence the choice of input into traditional modelling exercises; faced with the number and complexity of variables that can influence water quality and quantity, modellers need to make choices. While these choices will be based on professional judgment, it is almost impossible to be completely objective in the choice of variables or indicators used.

Better knowledge of human dimensions of modelling is therefore an important field that needs to be developed. Human factors influence input into and the understanding and use of models and human perceptions of the future, as well as human activities, are often incorporated into modelling. Data-rich models can be especially useful in predictions of possible futures and as such often constitute an input into policy-making by politicians, the business community, and other scientists. Scenarios attempt to construct general representations of possible alternative futures and they can be used to present the results of relatively complicated scientific modelling exercises in a manner that politicians and businessmen are able to understand (Gooch and Stålnacke, 2006). In organizational structures established ways of treating information influences the uptake of modelling outcomes and more knowledge is needed on the presentation and reception of modelling results in these contexts. We need to analyze the experiences gained in how models are used, and how they can be better tied to the needs of end-users, not only at the presentation stage, but also at the inception of the models. We also need to examine the ways that model results have been incorporated into decision-making and implementation procedures, and how their influence compares to the influence of other techniques. Are the models in use today the most suitable for the end-users? Do they provide the type of information that they ask for? How can we design models that take into account the needs of these users? Finally, we need to discuss if models can be used to present scientific information in an understandable way to a broader public, and to discuss the possibilities for the public to interact with modelling exercises through interactive techniques and through the development of on-line, fast connection technology which allows large groups of people to use computer programs for simulation without affecting the databases etc. that the information is based upon.

2.2.3. Case study III – The use of scenarios in the MANTRA-East Project

An example of the use of scenarios in water management is the MANTRA-East project (http://www.lwr.kth.se/Forskningsprojekt/mantra-east/index.htm). The MANTRA-East project ran from 2001–2004, and studied transboundary water management on the Estonian-Russian border, around Lake Peipsi. It had three main objectives. The first of these was to evaluate the applicability of the EU Water Framework Directive in the future EU border regions. This included an assessment of the state of eutrophication (e.g., ecological status) in lakes and river basins, as well as the development of strategic lake and river basin tools for source apportionment, retention, and time-trends in nutrient loads. The second objective was to develop methods to improve communication and utilization of scientific information. The third objective was to develop institutional mechanisms and policy instruments for decision making under conditions of transition and uncertainty, including institutional arrangements for public and stakeholder participation. The first objective was primarily the domain of the natural scientists in the project, the second objective was mostly managed by geographers and information specialists, and the third objective was the responsibility of the social scientists and policy analysts. The underlying rationale behind this model was that scientific information produced by the natural scientists was needed in the policy process, as well as knowledge of the policy process (decision-making and implementation) itself. This information needed to be communicated to the actors in the policy process, as well as to stakeholders and end-users. The project also needed input from the people affected, in other words, public participation.

The project utilized a combination of qualitative and quantitative scenarios, in which four possible futures for the region, in a 15-year perspective, were formulated. This part of the methodology analyzed possible developments in the population (growth or decrease), economy, agriculture, politics, administrative capacity etc. The results of this first stage were used as input into modelling exercises and simulations. Here the scenarios were quantified, and the difficult test of translating regional scenarios into local calculations of nitrogen and phosphorus loads etc. was managed. The results of the modelling, which indicated which factors (agricultural growth/decline, population growth/decline, level of urbanization) led to increases or decreases in nutrient loads. Finally, these results were used to formulate policy recommendations.

The choice of variables for the scenarios was made in cooperation with local stakeholders and policy-makers, and the preliminary scenarios were presented to and discussed by these groups. Here we can see that the choices made by the researchers were also influenced by local perceptions of the problems, an example of how human dimensions of two types, scientific and local, can be integrated.

From this short summary of the MANTRA-East project it can be seen that human dimensions of modelling were an important ingredient in the project. Human perceptions of what constituted the most important pressures on the environment and on water management were the starting point for the modellers. The choice of factors was dependent on the ways that the research team saw the interdependencies between different factors. The results of the modelling exercise were also translated into policy recommendations by the research team, and therefore also the result of human interaction and understanding of the modelling results.

2.2.4. Participatory modelling – What kind of input you are after?

We also need to discuss how models can be used to improve public participation, and to examine the potential that they have to present scientific information in an understandable way. It is also important to discuss the possibilities for the public to interact with modelling exercises through interactive techniques and also through the development of on-line, fast connection technology which allows large groups of people to use computer programs for simulation without affecting the databases etc. that the information is based upon.

Recent developments and paradigms in water management such as the WFD stipulate the involvement of the public and stakeholders in water management and the development of methods to do so. The preamble to the WFD acknowledges that: 'the success of the Directive relies on close cooperation and coherent action at Community, Member State and local level as well as on information, consultation and involvement of the public, including users' (EU, 2000). It can be seen here that the WFD distinguishes between three levels of participation, with an increasing degree of control over the outcomes of the process: these are 'information', 'consultation', and 'active involvement'. Information can be provided by means of brochures, briefings, press releases, etc. Here there is a one-way flow of information, downwards, from the initiators to the stakeholders. The stakeholders and public (or publics) do not themselves have the opportunity to give their opinion, as they only 'consume' the information that is offered to them. 'Consultation' can be considered a low level of participation. Examples of 'consultation' are interviews, opinion polls, and public hearings. The flow of information is still one-way, but this time upward, from the stakeholders and the public to the initiator. In both 'information' and 'consultation', there is no, or only limited, interaction within the stakeholder group, and between the water managers and the stakeholders. 'Active involvement', on the other hand, can be considered a higher degree of participation. 'Active involvement' can be organized by means of citizens' juries, planning cells, consensus conferences etc. For an overview of participatory methods, see: Van Asselt and Rijkens-Klomp (2002), Toth (2001), Creighton *et al.* (1998). The aims of participation can also be characterized as to inform, consult, involve, collaborate with, and to empower (IAP2, 2006). All of these forms of participation, however, are dependent on the provision of information to the public, and here we come up against the problem of presenting specialized knowledge to laymen in an understandable manner. Models are often used by natural scientists (hydrologists, biologists etc.) and might provide a suitable way of presenting results – if they can be tailored to the needs and experiences of the public.

However, if the public are to be involved in modelling in some way it is vital to determine the type of participation that we're looking for, as well as the possible outcome of that participation.

Elsewhere (Gooch and Huitema, forthcoming) it has been argued that participation in water management may have a number of aims. Participation may be used simply as a means of raising awareness of the issues at stake. A significant number of problems compete for attention on the public arena, and the involvement of the public in water management, in one way or another, may help to focus attention on the ones that water managers are most interested in. The use of models, if they are presented in a way that the public can comprehend, may serve an important purpose here, by increasing awareness of problems, possible solutions, and the interactions between them. A second goal of participation may be to make better decisions by including information about possible consequences of the intended measures that enable managers to formulate solutions. Here public participation can contribute to model building by providing new information on expectations, the acceptance of solutions, or on local conditions not known to the experts. Participatory exercises can demonstrate which priorities the public consider important, and the kind of choices they feel should be made, and these can be included in models. A third possible goal of participation is to increase the legitimacy of decision making, by reducing the distance between environmental managers and the public, and by engaging the public in deliberation about the decisions that need to be taken. If this is to succeed, however, the public must be able to feel that their input has been taken seriously. This can occur if the public are enabled to influence the input into models, as above, or when they are involved in the interpretation of model results. This last aspect may seem somewhat radical, as it is often a dearly held belief among some modellers that their models are objective and that the results can only be interpreted in one way. A fourth goal of public participation is to increase the accountability of decision making and responsiveness to the concerns of the public. Creating transparency and providing access to government actions and decisions is important, just as it is important for decision-makers to understand the public mind and find out what it is that people care about. Models can fulfil a function here too, if they are presented in a suitable manner. Finally, participation may result in collective learning. Stakeholders, government and scientific experts ideally enter into a dialogue and, by interaction and debate, learn how to collectively manage a natural resource (Pahl-Wostl, 2002) and get insight into conflicting views and interests. As a means of visualising problems, of exploring alternative course of action, and of presenting possible futures, models may be able to improve these processes.

2.2.5. Case study IV – Participatory modelling in South Africa

The following case study shows an application of participatory modelling in South-Africa (Farolfi and Rowntree, 2005).

According to the 1998 South-African National Water Law (NWL), the water users association (WUA) of the Kat River will be responsible in the long term for water resource management in the Kat catchment area. The present decision maker, the Department of Water Affairs and Forestry (DWAF) will mandate the WUA to allocate water rights and decide on water prices as soon as the WUA has formulated a Catchment Management Plan (CMP) and this plan has been validated by the DWAF.

The NWL requires the WUA to develop their CMP in a participatory way. Having no prior experience with participatory decision making, the WUA therefore looked for tools and approaches that would support the design and management of a process of dialogue and negotiation between representatives of very different types of water users. The WUA opted for the Companion Modelling approach (http://www.commod.org).

The following question was formulated as a starting point: *How to improve local management of water resources in Kat River basin, while allowing the various groups of users to express themselves?* The approach proceeded in four phases:

- Integrate the scientific knowledge that was available on the state and the uses of the water resources in the form of an agent-based model (KatAWARE).
- Share this representation of the Kat River system with the local actors (WUA) and improve it by incorporating local knowledge and critical feedback obtained with the initial model.

- Develop a role-playing game on the basis of the agent-based model and involve stakeholders in playing this game (two such game sessions were held).
- Revise the model and develop possible scenarios of water resource allocation using this new version of the agent-based model.

The approach led to a better understanding of the hydrological and socio-economic consequences of the adoption of alternative water allocation strategies in the Kat River basin. There was also intense interaction between actors, in particular during the role-playing meetings, which facilitated the dialogue and the discussion of water resource management scenarios between the participants.

The Kat WUA is currently in the phase of drafting the CMP, and the water allocation scenarios developed with the KatAWARE model are used as the basis for negotiation and for designing strategies to be proposed in the Plan. Although there still remain numerous stakeholder conflicts to be resolved, the participatory development and use of the KatAWARE model has helped to identify the decisions and interests, and to bring the stakeholders to the negotiation table.

The approach was fruitful politically as well as scientifically. In view of the positive results obtained in the Kat River basin, the Water Resource Commission (WRC) who finances the project has proposed to take the same approach in other South-African catchment areas in order to improve the process of negotiation and collective decision-making within the WUAs. The researchers involved could (after simplifying the context) compare the results of the role-playing games with a model of co-operative game theory (Dinar *et al.*, 2006) and verify to what extents the observed player behaviour was consistent with economic theory.

3. HUMAN DIMENSIONS IN THE USE OF MODELS

3.1. Introduction

The use of models to provide information and insight is a varied and variable process without deterministic outcomes as might be desired by the physical scientist stereotype depicted in Section 1.2. As we argued in Section 2.1, models are themselves the output of a modelling process in which the perceptions of the development team and the needs of the client or research funding agency will filter out and amplify different issues. This is more or less relevant depending on the type of phenomena being modelled – modelling mechanistic, physical processes is less prone to subjective bounding influences than modelling complex, potentially evolutionary social or socio-natural processes. However in both cases decisions are made about what will be included and what will be excluded from a model.

The use of models then adds an extra process of perception, an extra human dimension which should be appreciated both to ensure that the expectations of the developer regarding the impact of a model are not naive to the workings of social process (e.g., the biggest impact of a model may be to instigate learning) and to ensure that those who use models are aware of how issues of selectivity and amplification can influence model interpretation. This section seeks to provide an exploration of the ways in which social processes, human perceptions and behaviour can influence the understanding and use of models. The aim is not to rubbish models as inherently subjective and therefore uncertain and unusable, but rather to expose some of the issues involved in interacting with models to obtain information and insight with the aim of sensitizing both developers and users.

Considering the problems outlined above, how can we make models more understandable to decision-makers, stakeholders, and other presumptive users? Let us first look at various sources of model complexity and for each source the typical coping strategies.

The basic idea implied by Table 3.1 is that a model should provide answers to questions, and that human-computer interfaces should be oriented towards these questions. This suggests that the questions that will be posed need to be identified before and during model construction. In practice, however, these questions are difficult to elicit from the intended model audience, and when presented, the model will invariably raise new questions.

Table 3.1 also implies that the comprehensibility of a model is a function of the complexity of the model and the cognitive capacity of its user, but it does not address 'sense-making problems' (e.g., problem: variables, scales and/or visualizations that are meaningless for the audience; coping strategy: use analogies or metaphors). Are then the models in use today the most suitable for the end-users? Do they provide the type of information that they ask for, and how can we design models that take into account the needs of the users? Our own interpretation of 'linking models to the experiences of end-users' is that there are significant differences between the expectations of, for example, the different categories of users and producers sketched out in the prologue. It seems that much still needs to be done before models and users learn to interact successfully.

In this section we focus on the sense-making problem. Building on what we said about perceptions and frames in Section 2.1, we can distinguish different scales of social process. At the personal level the perceptions and values of the individual, as well as her/his level of education, will influence their understanding of model results. The frames of reference of the individual will determine which aspects of the model results will be accepted as 'true' and which will be refuted. Moving up a level, insights from social psychology lead us to expect that group pressures will lead to a distortion of the information provided by modelling, and that 'group-think' (Janis, 1972) may result in the refusal to accept certain results. Similarly, political and ethical considerations may also play a role. As models in water management are predominantly used in organizations, we also need to examine the ways that organizational or institutional forms influence the use of modelling outcomes. In organizational structures, institutional culture – which can be seen as a form of group influence, as well as established ways of treating information – influences the use of models, and prior experiences, as well as the different ways that information can be used, can lead to distortions of information flow. We must also distinguish between different scales or levels within an organization for the contribution and influence of models will be qualitatively different.

Table 3.1 Coping with model complexity

Source of model complexity	Coping strategy
Size = scope * level of detail	Hide aspects deemed irrelevant for the audience (reduce scope though selection, detail through aggregation)
Interaction effects between variables, non-linearity of relations	Visualize relations graphically; provide illustrative examples of the interactions that strongly co-determine system behaviour
Interdependencies between subsystems	Visualize subsystems and their relations using diagrams; identify the variables that subsystems have in common; develop scenarios that illustrate typical subsystem interactions
Micro-macro relations: (1) aggregation for equation-based models, (2) emergence for agent-based models	(1) provide views at different levels of aggregation with 'drill-down' options, (2) illustrate the underlying mechanism (if known)
Temporal relations: dynamic behaviour	Visualize behaviour (plot variables over time, show animated maps); identify typical patterns, highlight these and explain the underlying mechanism

Eason (1991), in exploring the use of information systems (of which models can be viewed a type), characterized organizations as socio-technical systems which have three levels with respect to the use of such systems:

- *Human-computer interaction.* This level focuses on the attitudes of users to using computer-based information systems, the factors which influence those attitudes and the ergonomics of computer input / output devices (e.g., mouse, keyboard, graphical user interface). Relevant factors identified include flexibility of the system (Bailey and Pearson, 1983; DeLone and MacLean, 1992; Eason, 1991), ease of use (Al-Ghatani and King, 1999; Bailey and Pearson, 1983; Davis, 1989), output information quality (Lee *et al.*, 2002; Nicolaou *et al.*, 1995), and experience of the user in using the system (DeLone and Maclean, 1992). However, understanding and predicting successful human-computer interaction has proved difficult because of the influence of many variables (Eason, 1991).
- *Task structure and environment.* Analyses at this level focus on the description of factors that determine the impact of computer-based information systems on individuals' work or on the performance of tasks carried out by groups of individuals. If systems have positive impacts on individuals work (e.g., higher task productivity), they are more likely to be used. Measures have narrowly concentrated on productivity impacts (Eason, 1991; Torkzadeh and Doll, 1999). However, Torkzadeh and Doll (1999) have attempted to conceive a more comprehensive framework that takes into account impacts on task innovation, management control and end-user satisfaction.
- *Socio-technical systems.* The majority of computer applications are implemented in organizations (Eason, 1991). Analyzing organizational usefulness is difficult as the influence of technology on an organization is a complex socio-technical phenomenon (Doll and Torkzadeh, 1998). The following factors need to be considered regarding processes of adoption and use: (i) attitudes of users towards the information system, (ii) impacts of the system on job effectiveness and efficiency, and (iii) organizational culture (Armstrong *et al.*, 2005; Cheney *et al.*, 1986; DeLone and MacLean, 1992; Doll and Torkzadeh, 1998; Eason, 1991; Limayem *et al.*, 2006; Raz and Goldberg, 2006; Winter *et al.*, 1995). It is at this level that Organizational Performance, Social Psychology, and Organizational Culture Theories (e.g., Cognitive Knowledge Identity and Adaptive Structuration theories) are particularly relevant (Limayen *et al.*, 2006; Raz and Goldberg, 2006).

Understanding the role of models at each level and how they, and the information they generate, are used will provide a greater insight into their impact on and usefulness to water policy and management. Unfortunately our understanding of the impact of models on the performance or structure of water management activities is not well developed. Indeed our understanding of what constitutes good performance in hydrologic, water quality, infrastructure, investment or demand management activities from the perspective of the organizations involved is similarly absent in the literature. Further, our understanding of the impact of models on less tangible aspects of water management like learning (reframing problems, changing priorities and values) is also not well developed.

Even without the psychological, social or organizational/cultural problems noted above, the reception of modelling results is often handicapped by insufficient understanding of communication techniques. Models can be viewed as a means of transferring scientific knowledge into practice and thereby contributing to the knowledge of those organizations that use such tools. Such a transfer of knowledge may have more immediate and direct performance benefits (e.g., using a new hydrological model to better predict flooding) but also longer term, indirect benefits in terms of learning and changing the aims and/or structure of organizational action (e.g., re-framing water management from supply to demand orientation and instituting demand management actions to improve resource availability). Many modellers believe that the results of modelling exercises will speak for themselves; however, this is often unfounded. We need therefore to discuss the use of models in organizations and the ways in which the presentation and communications of model results can be improved. We need also to discuss the experiences gained in how models are used, and how they can be better tied to the needs of end-users, not only at the presentation stage, but also at the inception of the models. In short we need to both better understand what it is that makes a model useful for people engaged in different types water management activity.

3.2. What makes a model useful?

We have already pointed out that providing relevant information in an understandable manner is a necessary condition for model use and usefulness. Beyond this criteria a range of authors have tried to determine the conditions under which information systems (and we treat models as a type of information system here) are more or less useful to organizations. These can be divided roughly into two groups:

- those which focus on understanding the processes by which information systems (IS) are adopted then used, or not adopted by individuals and organizations, and
- those which focus on processes of IS use and the impact IS have on individual (task) or organizational performance (measured in some way), individual or social learning or work structure.

The second group can be considered to be concerned with measurable impacts/outcomes of using models and the factors (model attributes, ways of using models, organizational factors like business processes and organizational values/norms) which influence impact/outcome, whilst the first group can be considered to focus on the subjective and perceptual aspects of model adoption and the factors (personal, group, technological, processual) which influence adoption/uptake.

As an example of the first group one of the models proposed is the Technology Acceptance Model (TAM), which establishes a direct relation between 'attitudes' or predisposition of users to utilize a tool (Al-Gahtani and King, 1999) and the perceived usefulness of that tool. Indeed up to 40% of variation in uptake and usage behaviour can be explained by the factors used within TAM – perceived ease of use and perceived relative improvement in user's job performance (so called perceived usefulness).

As an example of the second group, following the criteria presented by Adelman (1992), Winter *et al.* (1995) and Checkland and Holwell (1999), a useful IS would be one which provides relevant information and which contributes to improving the effectiveness and/or efficiency of organizational action:

- Models are meant to support people in organizations working together to take action through providing information representation, manipulation, storage and communication facilities. To do so it is clear that models must first and foremost provide information which is perceived as being relevant to action by members of the organization. The notion of relevance is therefore included in our definition of usefulness to highlight that useful models must provide information which satisfies the needs of people involved in taking organizational action.
- Useful models should contribute to improving the effectiveness of taking organizational action i.e., to improving the achievement of organizational

aims (Checkland and Scholes, 2004). More specific criteria for effectiveness will need to be defined by individual organizations and their actions. Hence there is a need for detailed empirical work beyond the scope of this paper to assess whether models contribute to improving effectiveness or not. Measures of effectiveness can alternatively be framed in terms of key performance indicators (KPIs).
- Last of all, useful models should also contribute to improving the efficiency of organizational action i.e., by reducing the effort, cost or resources required to produce a particular level of effectiveness (e.g., by facilitating communication, by speeding up data acquisition and analysis etc.) (Checkland and Scholes, 2004).

The largely business management oriented IS literature has a longer history of trying to understand the factors which influence both the adoption and use of models. In synthesizing the IS literature and interpreting it for water and land management, Diez and McIntosh (2009) provide the following table. It describes factors which influence the outcomes of different stages in the life cycle of an IS.

This section has shown that an evaluation of usefulness is obviously again a process strongly influenced by the perceptions and frames of those who define the usefulness of the specific model. Although it may be clear that a new model would significantly benefit (improve the performance of) a water management organization, the relationship between costs and benefits is crucial to adoption and therefore use and usefulness. Developers need to appreciate the nature and degree of change required by an organization to start using computer models or to start using new / different models. Using any new technology requires that the organizational system be reconfigured to some extent. Use of technology is a process of both shaping the technology to the need of the organization and shaping the organization to the demands of the technology. Staff may need to be trained, new IT put in place, work and business processes reconfigured etc. The cost of organizational change may be significant and difficult to estimate and this fact itself may pose a barrier to the adoption of models by water management organizations, particularly if the resulting performance gains are uncertain.

Table 3.2 Factors related to IS adoption, use and impact (Diez and McIntosh, 2009)

Process	Best predictor	Outcomes influenced	Type of relationship	Description of the relationship
Pre-implementation				
	User participation	User satisfaction	+	Involving users is likely to improve the quality of design decisions
		Success	+	Users are less likely to accept a IS unless they exert influence on its development
		Success	–	User participation can lead to some expectations that cannot be later satisfied by IS developers.
Implementation (individual and organizational scales considered together)				
	Behavioural intention	Actual system use	+	Individual use behaviour is driven by intention to use
	Computer experience	Adoption	+	More experience leads to more positive beliefs about technologies. Adoption is therefore more likely
	External pressure	Adoption and adoption intention	+	The adoption of IS is more likely if there is pressure from other organizations for work to be done or reported on in a particular way
	Information sources (external)	Adoption intention	+	IS are seen as devices for a better storage and management of information. Hence, their adoption is more likely if an organization has to handle large volumes of information from external sources

(*continued*)

Table 3.2 *Continued*

Process	Best predictor	Outcomes influenced	Type of relationship	Description of the relationship
	Perceived usefulness	Actual system use	+	The use of a IS is more likely if the user can foresee performance benefits as a result of IS use
	Professionalism (IS unit)	Adoption and adoption intention	+	The adoption of IS is more likely if IT departments are aware of the functionalities of the IS
	Subjective norms	Adoption	+	Perceived social peer pressure to adopt has been shown to significantly influence individual scale adoption outcomes
	System quality	Actual system use	+	Higher system quality tends to lead to more user satisfaction. The more satisfied the user is with the system the more he/she will be inclined to use it
	Top management support	Adoption	+	Locking in support from top managers has a strong positive influence on the adoption of IS. Typically developers must convince managers that IS will have positive performance impacts.
	User support	Adoption	+	The adoption of a IS is more likely if user support is high. User support in turn is positively influenced by satisfaction with the characteristics of the IS (e.g. ease of use, contribution to better performance)

(*continued*)

Table 3.2 *Continued*

Process	Best predictor	Outcomes influenced	Type of relationship	Description of the relationship
	User training	Actual system use	+	There is a direct relationship between training and use levels – more training also entails higher users' perceived relative advantage of the system, which in turn enhances the likelihood of IS use
Post-implementation				
	User satisfaction	Success	+	User satisfaction drives the continued use of IS

3.3. Models as tools to support action

Useful models can therefore be seen as technologies designed to help people undertake some kind of action through providing data processing and information generating facilities. Models are 'systems which serve' (Checkland and Holwell, 1999) and as such it is vitally important to appreciate the nature of the 'system which is served' by the use of models. It makes no sense to design models to nominally support water management without an understanding of organizational context – without an understanding of the 'system' which is to be supported through the information and facilities provided by a models. This system can be viewed as the set of linked activities undertaken by people within an organization in order to collectively take action. Organizational action is then defined as the whole set of linked activities involved in pursuing a collective purpose and the way in which the performance of such systems of action is assessed and change in action taken as a consequence (Checkland and Holwell, 1999).

In developing models we are seeking to provide tools which represent and process data and information in particular ways for use by people engaged in organizational action. In our case we are focussed on management action in the context of water from managing river water flows, to ground water quality, to improving wastewater treatment plant performance to informing asset investment and management to managing water demand. A key aim must be to improve organizational performance in either internal or external terms through improving the performance of individual tasks and their inter-relationships (dependencies). By internal here we mean measures of performance related to the achievement of internal effectiveness or efficiency objectives (e.g., cost measures, reducing staff turnover) and by external we mean measures of performance related to the achievement of external effectiveness objectives (e.g., attaining good ecological status for a river, reducing flood risk to households, receiving no complaints from customers).

For models to improve performance there must be a demand for the information they provide or knowledge that they can contribute to developing.

3.4. Data, information and knowledge

Before proceeding to explore the processes through which models, or more specifically model outputs, are interpreted and used as the basis for arguing for change or taking action in water management we will pause to clarify the meaning of three related concepts: data, information and knowledge. These concepts are often used interchangeably and the confusion created by their misuse it has been argued lie at the foundations of problems in developing useful information systems (Checkland and Holwell, 1999). It is clear that water management has a need for scientific data, information and knowledge and claims have been made that models provide a suitable means of satisfying these needs. So what are these concepts and how do they relate to model development and use?

"*Data is a set of discrete, objective events or facts*" (Davenport and Prusak, 2000). The veracity of stated facts can in principle be supported or refuted by gathering evidence. Unlike data however, information is subjective, resulting from selecting or paying heed to some of the data potentially available and, crucially, attributing meaning to it (Checkland and Holwell, 1999). Information is a set of selected facts (or data) which have been organized in some way and given meaning. For example data on soil type, precipitation and slope angle could be selected and combined to create information on erosion risk for a given area. However the process of selecting data and organizing it into meaningful information in an organization is subject to the cognitive filters of the people involved (what facts they perceive as relevant and why) and the effect of discourse between people on those cognitive filters (perception of relevance is both personally and socially influenced). As a consequence, people in organizations can attribute different meanings or allocate a different degree of importance to the same set of facts (data). Attributed meaning can then be shared by many organizations but might also be unique to each one (Alavi and Leidner, 2001). This emphasizes the idea that organizations obtain knowledge in accordance to the relative importance they give to specific pieces of data. Appreciating the influence of variations in meaning and importance is crucial with complex, messy issues like water management where conflict exists over the salient issues and how best to resolve them.

Knowledge is a fluid mix of framed experience, values, contextual information, and expert insight that provides a framework for evaluating and incorporating new information. It can involve different types such as know-how (knowledge of how to do something), know-what (knowledge of what to do under particular circumstances), know-why (knowledge of why a particular phenomenon is occurring) etc. (Davenport and Prusak, 2000). Knowledge is embedded in individuals or groups of people, knowledge is a larger and longer structure of meaningful facts or information (Checkland and Holwell, 1999; Alavi and Leidner, 2001). In other words, knowledge may be expected to have greater longevity than many items of information, which are only ephemerally meaningful and relevant (Davenport and Prusak, 2000). Knowledge and the processes of data selection and conversion into meaningful information will influence each other and change over time through learning.

Put in the necessarily social context that is organizational action, it is clear that models must provide information which is perceived as relevant and meaningful by the members of that organization. Relevance will be determined by the perceptions, frames and cognitive filters of the individuals involved and by the nature of action in which they participate – the specific activities involved, their overall collective purpose, and the criteria and processes by which performance is measured and corrective change undertaken. If a model provides information which is relevant (i.e., meaningful) to the tool developer but not to the intended user it is, we argue, unlikely to be used.

Model developers need therefore to be aware that specific models might be useful for people in one organization, but not for people in another, because of variations in what constitutes relevant information and variations in performance criteria between organizations (or even within an organization, between different human activity systems – different groups of people taking action). If a model does not provide relevant information support, it is unlikely to be used. This sounds trivial but lies at the core of developing useful models. Understanding what information is deemed relevant to a group of people engaged in collective action and why it is crucial to have effective model design and use.

3.5. The role of management and managers: how are models interpreted?

In the following two sections we take up the third level of model use, i.e., the use of models in organizations or management processes. Modelling involves flows of data and information and decision-makers need access to reliable and understandable information in order to make functional, logical and rational decisions. This flow of information usually takes place in institutional and organizational settings and involves managers and management practices. Here we need to distinguish between the terms 'organization' and 'institution', which are often seen as more or less synonymous. However, while both 'organization' and 'institution' can be used to refer to a group that works purposely together within a context of common rules, 'organization' suggests both a way of managing and a physical unit, while 'institution' can be seen as a set of norms and values. The ways that information is treated may not be optimal and often needs to be improved. In other words, the communication of modelling needs and results between the scientific community and policy-makers needs to be developed in agreed-upon contexts according to rational criteria. While the significance of institutionalized communication has long been accepted in studies of business administration, where communication and organization structures are often seen as intertwined (Gortner *et al.*, 1987), the importance of these institutional contexts in communication processes in public administrations has not always found such wide acceptance. This is surprising, as the input and results of modelling are managed by organizations and institutions, public or private, in which institutionalized communication plays a vital role.

3.6. The strategic use of models in organizations

Water policy and management organizations typically have clearly stated aims and objectives and objectives relating to aspects of their work e.g., improving or maintaining river water flows or quality, reducing leakage levels from water supply infrastructure. The strategic use of models refers to one way by which management and managers influence the interpretation of model results.

To capture better how this strategic use occurs we borrow from an institutional framework which builds on Pfeffer (1981) and was developed by Gooch. It is presented below. In what is of course a simplification of complex factors three dominant institutional models can be formulated; these models are the Rational Model, a normative ideal for many water managers, the Bureaucratic Model and the Political Model. Table 3.3 presents a number of aspects of the models such as goal and preferences, forms of power, decision processes, beliefs about action and consequences, forms of decision-making, ideologies, and organizational forms.

In this section we have identified the different ways that humans can use different types of models. We have claimed that water managers need to consider the characteristics of the different models, their usefulness for the task at hand, as well as the organizational context within which they are used.

Table 3.3 Institutional contexts in water management and the use of scientific information (Gooch, 2006)

Dimension	Rational	Bureaucratic	Political
Goals/preferences	Consistent and based on complete information	Determined by organizational rules	Pluralistic, formed by competing groups
Power	Centralized and transparent	Rule-determined and not always understandable by outsiders	Determined by coalitions that can change
Decision process	Orderly, attempts to mirror economic principles	Procedural, follows established patterns	Disorderly and incremental
Information use	Extensive and systematic	Used only when it fits into established rules and procedures	Used strategically by different coalitions
Beliefs about action-consequences	Based on known probability, predictable	Expected outcomes based on earlier experiences	Disagreement on outcomes between different coalitions
Decisions are	Perceived as value maximizing	The result of established routines	The result of bargaining between coalitions
Dominant ideology	Economic efficiency	Stability and predictability	Conflict as a creative force
Organization	Market-based, new public management	Hierarchical	Competing networks
Examples of theory	Institutional rational choice	Max Weber's theory of bureaucracy	Cultural theory/advocacy coalition framework
Theorists	Elinor Ostrom	Max Weber	Paul Sabatier

3.7. Political use of models

Closely connected with the strategic use of models in organizations is the way that models can also form part of the *political*, that is, the way that they can function not only as mediators of scientific theories between scientists and water managers, but also as a form of power in political decision-making. In the section above on the use of models in organizations and institutions, we claimed that models and other forms of information and communication can, and are, used strategically. However, water management is not only concerned with the administration and the *implementation* of decisions about water, but also to a high degree about the decision-making process itself. This involves decisions about 'who gets what, when and how; it involves the creation and exercise of power. Power should not however be seen as something static, but as '*a (concealed or misrepresented) effect, rather than power as a set of causes*' (Law, 1992). Actors gain power through their relations with others, and, as Foucault has pointed out, in many cases '*power is not in the hands of one person who can exercise it alone and totally over others, but is a result of those relationships*' (Foucault, 1980). Such relationships include the interaction of model-makers with model-users, as well as the competing uses of models by political actors. Through their explanatory and predictive abilities models represent a significant form of power and we must therefore also look at the ways that power is formed and exercised in the creation of models, and in their subsequent use in decision-making processes.

An example from one of the case studies in Section 2.2 can be used to elucidate this issue. In the short description of the MANTRA-East project the use of scenarios was introduced. While the scientific outcome of this process was very successful, the impact on policy-making was more uncertain. The MANTRA-East qualitative scenarios and subsequent statistical modelling demonstrated that the most effective strategy for water management around Lake Peipsi (between Estonia and Russia) would be investments to minimize the negative effects on water quality in the lake that modernisation of agricultural practices was expected to result in. At the time, Estonia had applied for membership in the European Union (EU), and the probable positive outcome of this application, and the introduction of the EU Common Agricultural Policy (CAP) was expected to result in increased use of artificial fertilizers and chemicals, in more efficient drainage of water-logged land, and in a faster run-off of nutrients (nitrogen and phosphorus). This increased run-off was expected to detrimentally influence the level of eutrophication of the lake. The MANTRA-East recommendations, based on the use of hydrological modelling, were that this run-off of nutrients should be minimized in order to protect the quality of the lake. However, the decision taken by politicians at the time was that available funds should be invested in water treatment plants for villages and small towns around the lake or situated on rivers in the lake basin. The official motivation was that the results of the statistical modelling were unsure; it may however have been the case that the investments necessary to limit the impacts of increased agricultural production would not be visible to large segments of the population. Investments in water treatment plants, however, have a high visibility impact. Here we have an example of how the legitimacy of models may be questioned, not necessarily because of the scientific reliability of the model, but because political realities (re-election and the consolidation of power) may influence the use of the model results in political processes and decision-making.

3.8. Case study V – Participatory stakeholder analysis for water management in Egypt

This case study illustrates the participatory use of a model to assess the political feasibility of different water management options in the process of the National Water Resources Plan (NWRP) in Egypt (Hermans *et al.*, 2002; Hermans and Bots, 2003). Having completed the analysis of the Egyptian water system, the NWRP project moved to the phase of drafting a policy document. A new national water policy should take the important values and positions of the different stakeholders into account, and its future implementation would require the support of a wide range of stakeholders.

The NWRP project management wished to intensify the discourse with the main stakeholders and discuss and clarify the support for, and opposition against, various policy alternatives. A metagame analysis (Howard, 1971, 1989) would reveal 'how the cards are dealt' and enable the NWRP analysts to identify promising ways to accommodate the different preferences of stakeholders. Participatory use of the metagame model would also help to make all the participants aware of their interdependencies, and increase their insight in their own interests and negotiation instruments.

The NWRP project team decided to select one specific, relatively simple water resources management issue that involved the main stakeholders in the Egyptian situation, perform an analysis of options on this issue to construct a metagame model, and then test the model with the team of NWRP engineers in Cairo. If this test session produced valuable results and the method proved to be efficient, then the next steps would be to scale up the metagame to include all the issues that are relevant to Egyptian water resources management and organize a participatory session with senior members of the NWRP stakeholder committees.

The test case involved five stakeholders: the Ministry of Water Resources and Irrigation (MWRI), the Ministry of Agriculture and Land Reclamation (MALR), the Egyptian Environmental Affairs Agency (EEAA), the Ministry of Health and Population (MHP), and the National Organization for Potable Water And Sewage Disposal (NOPWASD). Analysis of options revealed three generic scenarios: maximize drainage water re-use, limit its re-use, or stop its re-use altogether. The diagram in Figure 3.1 shows one of the results of the analysis: the strategic map that indicates which stakeholders have the power to move from one state (the result the actions of other stakeholders) to another.

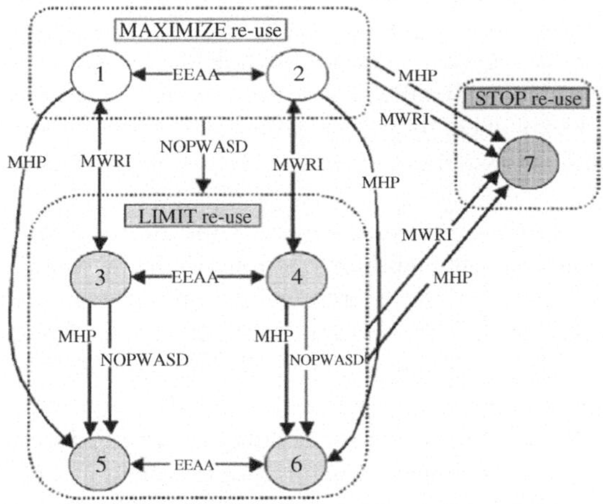

Figure 3.1 Strategic map showing basic states and unilateral moves of stakeholders

The metagame model was considered to be an adequate representation of reality, without being too complex. It was used as a basis for a three-hour simulation exercise to explore possible negotiation scenarios. The NWRP engineers in Cairo assumed the roles of the five stakeholders and enacted several rounds of policy negotiations, giving or withdrawing their support for different options until a feasible compromise was reached. This simulation exercise led to two important insights:

(1) Neither the National Organization for Potable Water and Sewage Disposal nor the Ministry of Health and Population exercised their power to make a unilateral move to a scenario they preferred over the

compromise. This suggests that the participants in the test session viewed the Ministry of Water Resources and Irrigation and the Ministry of Agriculture and Land Reclamation as more powerful stakeholders. Indeed, in reality, the Ministry of Agriculture and Land Reclamation is perceived to be a powerful entity in Egypt's public sector.

(2) The participants were creative in identifying variations on the predefined scenarios. The eventual compromise – optimize re-use – did not come as a total surprise. NWRP team members were already aware of the option of the Ministry of Health and Population to be flexible in applying the official water quality standards, allowing more re-use in areas where somewhat lower quality standards would also be sufficient to protect public health. In fact, this policy was already practiced. However, it was not included as an option in the metagame model, because officially it was not allowed. As some Egyptian team members explained, this option would never become official policy, and real politicians would not propose it even during a model use exercise.

The 'discovery' of the informal power of the Ministry of Agriculture and Land Reclamation and the unofficial option of the Ministry of Health and Population posed a problem for the developers of the metagame model. It would probably not be wise to make the power of the Ministry of Agriculture and Land Reclamation explicit by giving it a veto-option, just as it was not wise to explicitly model the pragmatic behaviour of the Ministry of Health and Population by including the option of 'bending the rules'. Informal power is by definition implicit and ambiguous. Capturing it in formal models is difficult, and when it is done it might well upset the real stakeholders.

The experiment confirmed that the metagame modelling approach helps to identify the more and less important stakeholders, and provides insight in preferences, interdependencies, and what trade-offs needed to be made. The simulation of negotiations between stakeholders inspired to identify compensation, resulting in new options for compromise. However, it also made clear that participatory use of a metagame model is problematic due to implicit and ambiguous power structures, stakeholders with hidden agendas, and various political and cultural sensitivities.

In the end, the NWRP project management decided not to use the metagame approach in a participatory setting with the project's real stakeholders. The problem of informal power and unofficial options may have played a role in this decision, but the main problem was the scope of the model. The senior officials that were part of NWRP's stakeholder committees would want to discuss the full range of water management issues in the region, and that would require a complete metagame model. Considering the effort needed to analyse the relatively simple case of drainage water re-use, the NWRP project management doubted whether the expected results of developing and using a full-scale metagame model would justify the huge effort this would require.

4. RECOMMENDATIONS

Belief systems, human attitudes, and collective behaviours have been presented in this Guidance Report as an integral part of both modelling and management objectives and strategies. Understanding human decision making has become of key importance with the advent of more open regulatory frameworks such as the WFD. We have attempted to cover a wide and poorly charted territory with this report and now need to bring our discussion to a close by making some practical recommendations regarding the development and use of models. To do so we will focus on providing guidance to three different types of actor involved in the development and use of models: the model commissioner (funding organization), the model developer (which may be a single organization or consortium of organizations), and the model user (which may be the same organization as or different from the commissioning organization). In providing guidance our ambition is not to provide a scientific or technical quality assurance procedure (such work can be found reviewed elsewhere, e.g., Refsgaard *et al.*, 2005), but rather to interpret the broader and underlying set of human dimension issues for different types users.

4.1. Guidance for model commissioners

The decision to commission a model is potentially one which involves committing significant budgetary resource and therefore should not be taken lightly. It will be crucial to minimize risk in terms of the model development process taking longer to complete than anticipated, delivering an unusable or unreliable tool or in the worst case, failing to deliver at all. The key recommendations from Caminiti (2004) as presented earlier, are to formulate a project brief prior to commissioning, and to engage with modelling experts independently from the model development team as a means of avoiding inappropriate mis-selling of particular approaches from model developers with vested interests or limited experience of the range of available approaches, and of ensuring the project brief is delivered to.

The downside of a detailed project brief is that it completely fixes the purpose and content of a model, discouraging the uptake of important new insights obtained during the process of model development. Especially for interdisciplinary projects, it is recommended to allocate significant additional time and resources specifically for frame reflection and negotiation in the beginning of the project. Letting the model developers specify up-front how they will organize their model bounding process, and asking for deliverables that document this process, their bounding decisions and deliberations will mitigate their tendency to focus on the model's content and the subsequent risk of poorly integrated results.

4.2. Guidance for model developers

Models involve human actions and are used by people. This simple fact means that models must take into account human dimensions in a number of ways. First, human behaviour is seldom easy to predict, and models that attempt to incorporate deterministic aspects of individuals and societies will invariably get things wrong. Individuals and groups behave in irritating ways, they often fail to follow what may be seen by model developers as logical patterns. The involvement of social scientists may partly help to understand possible deviances, but social scientists also often get things wrong. The alternative is to attempt to incorporate alternative behaviour patterns, even those that do not seem especially likely, into the models.

Second, for water resource management issues modelling teams will have to include a specific range of disciplinary backgrounds. Each discipline has its own preferred ways of representing social and/or physical phenomena. It is important to realize that model development is cognitively a very demanding process. As modellers, we tend to forget that what we see as meaningful and elegant representations are not easily understood and appreciated by those who have not gone through the many steps of thought that have led to these representations. Spending some precious time on developing didactical materials that will help others understand the model will pay off in the end.

Third, model development takes place in a complex social context. The roles and job responsibilities of the people in this context will not always be transparent, and model commissioners, model developers and model users each may have different perceptions and expectations of the others. Without making the effort to understand these perceptions and the underlying frames, the other effort put into a project is likely to be less effective. As model commissioners are recommended to fund frame reflection and negotiation processes, model developers are recommended to ask for such funds and plan for these processes. Here, it is important to acknowledge the expertise of professional process facilitators.

4.3. Guidance for model users

Concerning human dimensions of models it is important for model users to be aware of the choices that have been made during model development, and especially how and why the choice of indicators, factors and variables has been made. While it is probably impossible to develop a model without these influences, especially as model development involves attempts to simplify complex conditions, it can help to identify the kind of choices that have been made, and to take into consideration the possible effects of those choices. This does not mean that models can only be used by persons who have fully comprehended the modelling process. If models are presented as finalized tools, e.g., to learn about the system simulated, the users have to preserve a critical attitude towards the model and demand explanations if the results seem to be at odds with their experiences or knowledge.

5. CONCLUSION

In this Guidance Report we have attempted to identify some of the most important aspects of human dimensions of modelling. We have looked at the influences of individual, group and societal beliefs, and at how selective perceptions of problems and solutions can influence model development and use. Transparency and well-defined process rules that structure interaction are the key to successful model development and use. While it is not possible in a report such as this to comprehensively deal with such a complex subject, we would like to finish off by encouraging colleagues on different sides of the model divides to take the step to integrate their knowledge and experiences. Our experience is that statisticians, policy-makers, social scientists, natural scientists and all others involved in the fascinating business of modelling have much to give and much to learn from each other. And after all, all our models are created and used by humans, so the human dimension cannot be isolated from all the others.

6. REFERENCES

Adelman, L. (1992). *Evaluating Decision Support and Expert Systems.* New York, John Wiley and Sons, 232 pp.

Alavi, M. and Leidner, D.E. (2001). Knowledge management systems: issues, challenges, and benefits. *Communication of the Association for Information Systems* **1**, 1–37.

Al-Gahtani, S.S. and King, M. (1999). Attitudes, satisfaction and usage: factors contributing to each in the acceptance of information technology. *Behaviour and Information Technology* **18**(4), 277–297.

Anderson, R.J. (1980). *Cognitive Psychology and it's Implications.* San Francisco, Freeman.

Argent, R. and Houghton, B. (2001). Land and water resources model integration: software engineering and beyond. *Advances in Environmental Research* **5**, 351–359.

Armstrong, B., Fogarty, G., Dingsdag, D. and Dimbleby, J. (2005). Validation of a computer user satisfaction questionnaire to measure IS success in small business. *Journal of Research and Practice in Information Technology* **37**(1), 27–42.

Bailey, J.E. and Pearson, S.W. (1983). Development of a tool for measuring and analyzing computer user satisfaction. *Management Science* **29**, 530–545.

Barreteau, O. (2003). The joint use of role-playing games and models regarding negotiation processes: characterization of associations. *Journal of Artificial Societies and Social Simulation,* **6**(2). http://jasss.soc.surrey.ac.uk/6/2/3.html.

Beers, P.J. and Bots, P.W.G. (2007). Eliciting conceptual models to support interdisciplinary research. *Proceedings of the 40th Annual Hawaii International Conference on System Sciences.* Los Alamitos, IEEE Computer Society Press.

Bonabeau, E. (2002). Agent-based modeling: methods and techniques for simulating human systems. *Proceedings of the National Academy of Sciences of the United States of America* **99**(10), 7280–7287.

Bots, P.W.G. (2007a). Benchmarking in Dutch urban water management: An assessment. In: Pahl-Wostl, C. (Ed.) *Proceedings of the International Conference on Adaptive & Integrated Water Management (CAIWA).* Berlin, Springer.

Bots, P.W.G. (2007b). Analysis of multi-actor policy contexts using perception graphs. In: *Proceedings of the IEEE/WIC/ACM International Conference on Intelligent Agent Technology.* Los Alamitos, IEEE Computer Society Press.

Bots, P.W.G. and De Bruijn, J.A. (2002). Effective knowledge management in professional organizations: Going by the rules. *Proceedings of the 35th Hawaii International Conference on System Sciences.* Los Alamitos, IEEE Computer Society Press. Retrieved 2007.03.09, http://www.hicss.hawaii.edu/HICSS_35/HICSSpapers/PDFdocuments/DDOML10.pdf

Brugnach, M. and Pahl-Wostl, C. (2008). A broadened view on the role for models in natural resource management: Implications for model development. In: Claudia Pahl-Wostl, Pavel Kabat and Jörn Möltgen (Eds.) *Adaptive and Integrated Water Management Coping with Complexity and Uncertainty.* Springer, Berlin.

Callon, M. (1986). Some elements of a sociology of translation. Domesticating the scallops and fishermen of St Brieuc Bays. In: Law, J. (Ed.) *Power, Action, Belief: a New Sociology of Knowledge?* London, England, Routledge and Kegan Paul, pp. 196–233.

Caminiti, J.E. (2004). Catchment modelling – a resource manager's perspective. *Environmental Modelling and Software* **19**, 991–997.

Carpenter, S., Broock, W. and Hanson, P. (1999). *Ecological and Social Dynamics in Simple Models of Ecosystem Management*, http://www.consecol.org/vol3/iss2/art4

Checkland, P. (1999). *Soft Systems Methodology: a 30-year retrospective.* Chichester, England, John Wiley & Sons.

Checkland, P. and Holwell, S. (1999). *Information, Systems and Information Systems: Making Sense of the Field.* Chichester, England, John Wiley & Sons.

Checkland, P. and Scholes, P. (1990). *Soft Systems Methodology in Action.* Chichester, England, John Wiley & Sons.

Cheney, P.H., Mann, R.I. and Amoroso, D.L. (1986). Organizational factors affecting the success of end-user computing. *Journal of Management Information Systems* **3**(1), 65–80.

Clark, H.H. and Schaefer, E.F. (1989). Contributing to discourse. *Cognitive Science,* **13**, 259–294.

Creighton, J.L., Dunning, C.M. and Delli Priscoli, J. (Eds.) (1998). *Public Involvement and Dispute Resolution: A Reader on the Second Decade of Experience at the Institute of Water Resources.* U.S. Army Corps of Engineers. Fort Belvoir, Institute of Water Resources.

Crutzen, P.J. and Stoermer, E.F. (2000). The "Anthropocene". *Global Change Newsletter,* **41**, 12–13.

Dake, K. (1992). Cultural theory and economic development: a proposal for a cultural audit. *Conference on Risk and Values.*

Davenport, T.H. and Prusak, L. (2000). *Working Knowledge: How Organizations Manage what they Know.* Boston, MA, Harvard Business School Press, 224 pp.

Davis, F.D. (1989). Perceived usefulness, perceived ease of use, and user acceptance of information technology. *MIS Quarterly* **13**(3), 3 18–340.

De Jouvenel, H. (2000). A brief methodological guide to scenario building. *Technological Forecasting and Social Change* **65**(1), 37–48.

DeLone, W.H. and McLean, E.R. (1992). Information systems success: the quest for the dependent variable. *Information Systems Research* 3(1), 60–94.

Demeritt, D. (2002). What is the 'social construction of nature'? A typology and sympathetic critique. *Progress in Human Geography* **26**(6), 767–790.

Diez, E. and McIntosh, B.S. (2007). Assessing the factors which make Decision Support Tools useful for water and land management. In: Proceedings *7th International IWA Symposium on Systems Analysis and Integrated Assessment in Water Management (WATERMATEX 2007),* May 7–9, 2007, Washington DC, USA

Diez, E. and McIntosh, B.S. (2009). A review of the factors which influence the use and usefulness of information systems. *Environmental Modelling and Software* **24**(5), 588–602.

Dinar, A., Farolfi, S., Patrone, F. and Rowntree, K. (2006). To negotiate or to game theorize: Negotiation vs. game theory outcomes for water allocation problems in the Kat basin, South Africa. Presented at the *6th Meeting on Game Theory and Practice,* Zaragoza, Spain 10–12 July 2006.

Doll, W.J. and Torkzadeh, G. (1998). Developing a multidimensional measure of system-use in an organizational context. *Information & Management* **3**(4), 171–185.

Donaldson, T. and Preston, L. (1995). The stakeholder theory of the corporation: Concepts, evidence and implications. *Academy of Management Review* **20**(1), 65–91.

Dorge, J. and Windolf, J. (2003). Implementation of the water framework directive – Can we use models as a tools in integrated river basin management? *International Journal of River Basin Management* **1**(2), 165–171.

Douglas, M. (1966). *Purity and Danger.* London, England, Routledge & Keagan Paul.

Douglas, M. (Ed.) (1982). *Essays in the Sociology of Perception.* London, England, Routledge & Kegan Paul.

Downward, S.R. and Taylor, R. (2007). An assessment of Spain's Programa AGUA and its implications for sustainable water management in the province of Almería, southeast Spain. *Journal of Environmental Management,* **82**, 277–289.

Eason, K.D. (1991). Ergonomic perspectives on advances in human-computer interaction. *Ergonomics* **34**(6), 721–741.

EC (2003). Economics and the Environment – The Implementation Challenge of the Water Framework Directive. Annexes. *Common Implementation Strategy (CIS) Guidance Documents No. 1.* Retrieved 2007-09-01, from http://circa.europa.eu/Public/irc/env/wfd/library?l = /framework_directive/guidance_documents/guidance snos1seconom

EU (2000). *Community Legislation.* Directive 2000/60/EC of the European Parliament and of the Council of 23 October 2000 establishing a framework for Community action in the field of water policy. Retrieved 2001-06-20, from http://europa.eu.int/eur-lex/en/lif/dat/2000/en_300L0060.html

Farolfi, S. and Rowntree, K. (2005). Accompanying local stakeholders in negotiation processes related to water allocation through simulation models and role-playing games: An experience from South Africa. *EMPOWERS Regional Symposium: End-Users Ownership and Involvement in IWRM,* 13–17 November 2005, Cairo, Egypt.

Festinger, L. (1957). *A Theory of Cognitive Dissonance.* Stanford, CA, Stanford University Press.

Fishbein, M. and Ajzen, I. (1975). *Belief, Attitude, Intentions and Behavior: An Introduction to Theory and Research.* Reading, MA, Addison-Wesley.

Freeman, R.E. (1984). *Strategic Management: A Stakeholder Approach.* Boston, MA, Pitman.

Gato, S., Jayasuriya, N. and Roberts, P. (2007). Temperature and rainfall thresholds for base use urban water demand modelling. *Journal of Hydrology* **337**(3–4), 364–376.

Giordano, R., Passarella, G., Uricchio, V.F. and Vurro, M. (2007). Integrating conflict analysis and consensus reaching in a decision support system for water resource management. *Journal of Environmental Management* **84**, 213–228.

Gooch, G.D. (2006). Actor network theory as a tool for analyses of multi-level water governance. *International Workshop on Governance and the Global Water System: Institutions, Actors, Scales of Water Management Facing the Challenges of Global Change.* Bonn, Germany, Global Water System Project.

Gooch, G.D. and Stålnacke, P. (Eds.) (2006). *Integrated Transboundary Water Management in Theory and Practice: Experiences from the New EU Eastern borders.* London, England, IWA Publishing.

Gortner, H.F., Mahler, J. and Nicholson, J. (1987). *Organization Theory: A Public Perspective.* Chicago, IL, The Dorsey Press.

Gough, C., Castells, N. and Funtowicz, S. (1998). Integrated assessment: an emerging methodology for complex issues. *Environmental Modelling and Assessment* **3**, 19–29.

Gray, D.B. (1985). *Ecological Beliefs and Behaviors: Assessment and Change.* Westport, Greenwood Press.

Greeuw, S.C.H., Van Asselt, M.B.A., Grosskurth, J., Storms, C.A.M.H., Rijkens-Klomp, N., Rothman, D.S. and Rotmans, J. (2000). *Cloudy Crystal Balls – An assessment of recent European and global scenario studies and models.* Copenhagen, Denmark, European Environment Agency.

Guyot, P. and Honiden, S. (2006). Agent-Based Participatory Simulations: Merging Multi-Agent Systems and Role-Playing Games. *Journal of Artificial Societies and Social Simulation* **9**(4). http://jasss.soc.surrey.ac.uk/9/4/8.html

Haggard, S. and Simmons, B.A. (1987). Theories of international regimes. *International Organization* **41**(3), 491–517.

Hare, M. and Pahl-Wostl, C. (2002). Stakeholder categorisation in participatory integrated assessment processes. *Integrated Assessment* **3**(1), 50–62.

Heberlein, T.A. and Black, J.S. (1981). Cognitive consistency and environmental action. *Environment and Behavior* **13**(6), 717–734.

Hermans, L.M. (2005). *Actor Analysis for Water Resources Management – Putting the Promise into Practice.* Doctoral dissertation. Delft, The Netherlands, Delft University of Technology.

Hermans, L.M. and Bots, P.W.G. (2003). Metagames: Exploring participatory stakeholder analysis for water management in Egypt. In: Mayer, I.S., Veeneman, W. (Eds.) *Gaming and Simulation in Infrastructure Development.* Delft, The Netherlands, Lemma Publishers.

Hermans, L.M., El Masry, N. and Sadek, T.M. (2002). Linking actors and models for water policy development in Egypt: analyzing actors and their options. *Knowledge, Technology and Policy* **14**(4), 57–74.

Howard, N. (1971). *Paradoxes of Rationality: Theory of Metagames and Political Behavior.* Cambridge, MA, The MIT Press.

Howard, N. (1989). The manager as politician and general: the metagame approach to analysing cooperation and conflict. In: Rosenhead, J. (Ed.) *Rational Analysis for a Problematic World: Problem Structuring Methods for Complexity, Uncertainty and Conflict.* Chichester, England, John Wiley & Sons.

IAP2 (2006). *IAP2 Public Participation Spectrum.* Denver, CO, International Association for Public Participation. http://www.iap2.org/associations/4748/files/Spectrum.pdf

Jakeman, A.J., Letcher, R.A. and Norton, J.P. (2006). Ten iterative steps in development and evaluation of environmental models, *Environmental Modelling and Software,* **21**, 602–614.

Janis, I.L. (1972). *Victims of Groupthink.* Boston, MA, Houghton Mifflin Company.

Jeffrey, P. and McIntosh, B.S. (2006). Description, diagnosis, prescription: Evaluating the application of co-evolutionary models to sustainability and natural resource management. *Environmental Conservation* **33**(4), 281–293.

Karstens, S.A.M., Bots, P.W.G. and Slinger, J.H. (2007). Spatial boundary choice and the views of different actors. *Environmental Impact Assessment Review* **27**(5), 386–407.

Keen, P.G.W. (1981). Value analysis: justifying decision support systems. *MIS Quarterly,* **5**(1), 1–15.

Latour, B. (1993). *We Have Never Been Modern.* Hemel Hempstead, UK, Harvester Wheatsheaf.

Latour, B. (1999). *Pandora's Hope: Essays on the Reality of Science Studies.* Cambridge, MA/London, Harvard University Press.

Latour, B. (1999). On recalling ANT. In: Law, J., Hassard, J. (Eds.) *Actor Network Theory and After.* Oxford, England, Blackwell Publishers, pp. 15–25.

Lee, Y.W., Strong, D.M., Kahn, B.K. and Wang, R.Y. (2002). AIMQ: a methodology for information quality assessment. *Information & Management* **40**, 133–146.

Limayem, M., Banerjee, P. and Ma, L. (2006). Impact of GDSS: Opening the black box. *Decision Support Systems* **42**(2), 945–957.

López-Paredes, A., Saurí, D. and Galán, J.M. (2005). Urban water management with artificial societies of agents: the FIRMABAR simulator. *Simulation* **81**(3), 189–199.

March, J.G. (1978). Bounded rationality, ambiguity, and the engineering of choice. *Bell Journal of Economics* **9**, 587–608.

McIntosh, B.S., Jeffrey, P., Lemon, M. and Winder, N. (2005). On the design of computer-based models for integrated environmental science. *Environmental Management* **35**, 741–752.

Mitsch, W.J. and Jørgensen, S.E. (1989). *Ecological Engineering: An Introduction to Ecotechnology.* New York, John Wiley & Sons, 472 pp.

Nickles, M., Rovatsos, M. and Weiß, G. (Eds.) (2004). Agents and computational autonomy – potential, risks, and solutions. *Lecture Notes in Computer Science* **2969**, 83–94.

Nicolaou, A., Masoner, M.M. and Welker, R.B. (1995). Intent to enhance information systems as a function of system success. *Journal of Information Systems* **9**(2), 93–108.

Nicolson, C.R., Starfield, A.M., Kofinas, G.P. and Kruse, J.A. (2002). Ten heuristics for interdisciplinary modeling projects. *Ecosystems* **5**, 376–384.

Ogden, P.E. and Hall, R. (2004). The second demographic transition, new household forms and the urban population of France during the 1990s. *Transactions of the Institute of British Geographers* **29**, 88–105.

Oxley, T., McIntosh, B.S., Winder, N., Mulligan, M. and Engelen, G. (2004). Integrated modelling & decision support tools: a Mediterranean example. *Environmental Modelling and Software* **19**, 999–1010.

Pahl-Wostl, C. (2002). Towards sustainability in the water sector: the importance of human actors and processes of social learning. *Aquatic Sciences* **64**, 394–411.

Pahl-Wostl, C. (2007). The implications of complexity for integrated resources management. *Environmental Modelling and Software,* **22**, 561–569.

Pahl-Wostl, C. (in press). Participation in environmental scenarios. In: Alcamo, J. (Ed.), *Scenarios of the Future – the Future of Scenarios.* Berlin, Germany, Springer Verlag.

Parker, D.C., Berger, T. and Manson, S.M. (Eds.) (2001). *Agent-Based Models of Land-Use and Land-Cover Change.* Report and Review of an International Workshop, October 4–7, 2001, Irvine, CA, USA. Retrieved 2007.09.02, http://www.globalland project.org/LUCC_No_6.pdf.

Parker, P., Letcher, R., Jakeman, A., Beck, M.B., Harris, G., Argent, R.M., Hare, M., Pahl-Wostl, C., Voinov, A., Janssen, M., Sullivan, P., Scoccimarro, M., Friend, A., Sonnenshein, M., Barker, D., Matejicek, L., Odulaja, D., Deadman, P., Lim, K., Larocque, G., Tarikhi, P., Fletcher, C., Put, A., Maxwell, T., Charles, A., Breeze, H., Nakatani, N., Mudgal, S., Naito, W., Osidele, O., Eriksson, I., Kautsky, U., Kautsky, E., Naeslund, B., Kumblad, L., Park, R., Maltagliati, S., Girardin, P., Rizzoli, A., Mauriello, D., Hoch, R., Pelletier, D., Reilly, J. Olafsdottir, J. and Bin, S. (2002). Progress in integrated assessment and modelling. *Environmental Modelling and Software* **17**, 209–217.

Parunak, H.V.D., Savit, R. and Riolo, R.L. (1998). Agent-based modeling vs. equation-based modeling: a case study and users' guide. *Lecture Notes in Artificial Intelligence* **1534**, 10–25.

Petty, R.E., Cacioppo, J.T. and Goldman, R. (1981). Personal involvement as a determinant of argument-based persuasion. *Journal of Personality and Social Psychology* **41**(5), 847–855.

Pfeffer, J. (1981). *Power in Organizations.* Marshfield, MA, Pitman Publishing.

Poch, M., Comas, J., Rodriguez-Road, I., Sanchez-Marre, M. and Cortes, U. (2005). Designing and building real environmental decision support systems. *Environmental Modelling and Software* **19**, 857–873.

Porter, M. (1985). *Competitive Advantage.* New York, The Free Press.

Putnam, R.D. (1988). Diplomacy and domestic politics: the logic of two-level games. *International Organization* **42**, 427–460.

Radford, K.J. (1984). Simulating involvement in complex situations. *Omega* **12**(2), 125–130.

Raz, O. and Goldberg, A. (2006). How to make IT work: Cognitive perspectives for better information technologies performance. *European Management Journal* **24**(2–3), 199–205.

Refsgaard, J.C., Henriksen, H.J., Harrar, W.G., Scholten, H. and Kassahun, A. (2005). Quality assurance in model based water management – Review of existing practice and outline of new approaches. *Environmental Modelling and Software* **20**, 1201–1215.

Rekolainen, S., Kämäri, J. and Hiltunen, M. (2003). A conceptual framework for identifying the need and the role of models in the implementation of the Water Framework Directive. *International Journal of River Basin Management* **1**(4), 347–352.

Rein, M. and Schön, D. (1991). Frame-reflective policy discourse. In: Wagner, P., Hirschon Weiss, C., Wittrock, B., Wollman, H. (Eds.) *Social Sciences and Modern States, National Experiences and Theoretical Crossroads.* Cambridge, England, University Press.

Richards, A.R. and Singh, N. (1997). Two level negotiations in bargaining over water. In: Parthasarathy, T., *et al.* (Eds.) *Game Theoretical Applications to Economics and Operations Research.* Boston, MA, Kluwer Academic Publishers.

Rittel, H.J. and Webber, M.W. (1973). Dilemmas in a general theory of planning. *Policy Sciences* **4**, 155–169.

Rokeach, M. (1960). *The Open and Closed Mind.* New York, Basic Books.

Rokeach, M. (1973). *The Nature of Human Values.* New York, The Free Press.

Sabatier, P.A. (1988). An advocacy coalition framework of policy change and the role of policy-oriented learning therein. *Policy Sciences* **21**(2), 129–168.

Schlumpf, C., Pahl-Wostl, C., Schönborn, A., Jaeger, C.C. and Imboden, D.M. (2001). IMPACTS: An information tool for citizens to assess impacts of climate change from a regional perspective. *Climate Change* **51**, 199–241.

Scott, J. (1991). *Social Network Analysis. A Handbook.* London, England, Sage.

Smouts, M.-C. (1998). The proper use of governance in international relations. *International Social Science Journal* **50**(1), 8 1–89.

Sterman, J.D. (2000). *Business Dynamics: Systems Thinking and Modelling for a Complex World.* Boston, MA/London, England, McGraw-Hill/Irwin.

Stoker, G. (1998). Governance as theory: five propositions. *International Social Science Journal* **50**(1), 17–28.

Schwarz, M. and Thompson, M. (1990). *Divided We Stand.* Hemel Hempstead, UK, Harvester Wheatsheaf.

Thomas, W.I. (1966). *W.I. Thomas on Social Organization and Social Personality* In: M. Janowitz (Ed.). Chicago, IL, University of Chicago Press.

Thompson, M., Ellis, R.J., Ellis, R. and Wildavsky, A.B. (1990). *Cultural Theory.* Boulder, SF, Westview Press.

Torkzadeh, T. and Doll, W.J. (1999). The development of a tool for measuring the perceived impact of information technology on work. *Omega* **27**(3), 327–339.

Toth, F. (2001). *Participatory Integrated Assessment Methods. An Assessment of Their Usefulness to the European Environmental Agency.* Copenhagen, Denmark, European Environmental Agency.

Van Asselt, M. and Rijkens-Klomp, N. (2002). A look in the mirror: Reflection on participation in integrated assessment from a methodological perspective. *Global Environmental Change* **12**, 167–184.

Van der Heijden, K. (1996). *Scenarios, the Art of Strategic Conversation.* Chichester, England, John Wiley & Sons.

Van Eeten, M.J.G. (1999). *Dialogues of the Deaf: Defining New Agendas for Environmental Deadlocks.* Delft, The Netherlands, Eburon Publishers.

Vanrolleghem, P.A. (in press) *Decision Support for WFD implementation – Water Framework Directive Series, Volume 3.* IWA Publishing, London, UK.

Vennix, J.A.M. (1996). *Group Model Building: Facilitating Team Learning using System Dynamics.* Chichester, England, John Wiley & Sons.

Vickers, G. (1965). *The Art of Judgement, A Study of Policy Making.* London, England, Chapman & Hall.

Weick, K.E. (1979). *The Social Psychology of Organizing.* Reading, MA, Addison-Wesley.

Willemain, T.R. (1994). Insights on modelling from a dozen experts. *Operations Research* **42**, 213–222.

Willemain, T.R. (1995). Model formulation: what experts think about and when. *Operations Research* **43**(6), 916–932.

Williamson, P., Mitchell, G. and McDonald, A.T. (2002). Domestic water demand forecasting: A static microsimulation approach. *Water and Environment Journal,* **16**(4), 243–248.

Winter, M.C., Brown, D.H. and Checkland, P.B. (1995). A role for soft systems methodology in information systems development. *European Journal of Information Systems* **4**, 130–142.

Young, O.R. and Osherenko, G. (1993). Testing theories of regime formation: Findings from a large collaborative research project. In: Rittberger, V., Mayer, P. (Eds.) *Regime Theory and International Relations.* Oxford, England, Clarendon Press, pp. 223–251.

Guidance Report II.5

Economic methods, models and instruments for the Water Framework Directive

Roy Brouwer, David Barton and Frans Oosterhuis

Abstract

This Guidance Report aims to provide an overview of the key issues in the economic policy analysis underlying the WFD. The report's target group is economic practitioners, water policy advisors and consultants with some familiarity of economics working on WFD implementation. Some knowledge and understanding of economics is expected to be necessary when reading the report.

The following five key issues were identified, described and elaborated in the paper: integrated river basin information systems, the selection of a cost-effective program of measures, the assessment of disproportionate costs, the estimation of environmental and resource costs, and the design of incentive compatible water pricing mechanisms. International examples are used to illustrate the issues, but practical examples directly related to WFD implementation besides the available article 5 descriptions are still scarce. Following the discussion of these key issues, economic textbook overviews were given of available methods, tools and instruments to address the economic policy analysis challenges in the context of the WFD. These include cost-effectiveness analysis, cost-benefit analysis, multi-criteria analysis, integrated hydro-economic modelling and economic instruments.

Keywords

Cost-benefit Analysis, Cost-Effectiveness Analysis, Economic Instruments, Economic Policy Analysis, Multi-criteria analysis, Water Pricing

1. INTRODUCTION

This Guidance Report presents some of the key issues in the economic analysis underlying the European Water Framework Directive (WFD) and provides an overview of the available economic methods and models to address these issues. The WFD is one of the first European directives in the domain of water, where economics is an integral part of the decision-making process surrounding its implementation in Member States (MS). The implementation process has been an important learning process so far for both experts and policy makers. Many decision issues have been clarified in the past years, but equally as many policy and research questions related to the development of the first integrated river basin management plans in 2009 remain open.

The identified key issues for the future implementation of the WFD across European river basins include:

- integrated water-economic river basin information systems and indicators;
- selection of cost-effective programs of measures to reach the WFD objectives;
- definition and assessment of disproportionate costs to reach the WFD objectives;
- cost recovery of water services including environmental and resource costs;
- design and evaluation of incentive water pricing instruments.

These issues surrounding the development of integrated river basin management plans are inherently multi-disciplinary in nature, requiring the input from hydrology, civil engineering, ecology and economics. For this interdisciplinary approach to be successful, the involvement of the users of the information (policy makers) is paramount (Brouwer *et al.*, 2003). The issues have partly been identified based on the economic requirements in the WFD and partly on specific national and European economic research programs in support of specific unresolved issues in the implementation of the WFD.

The report's main objective is to highlight the key issues in the economic analysis in the WFD towards the first integrated river basin management plans, and to present an overview of available economic methods and tools to address these issues. The paper is written for economic practitioners, water policy advisors and consultants working on WFD implementation, who are familiar with economics.

The key issues and tools and methods are illustrated using examples largely taken from own work in this field and are therefore inevitably biased towards North European case studies. However, in view of the authors' participation in pan-European networks (including Common Implementation Strategy working groups) surrounding the WFD implementation, the key issues raised in this paper are believed to be wider applicable. Another important disclaimer is that the examples do not necessarily reflect state-of-the-art practices in the actual

WFD implementation across MS. Political-economic conditions are highly influential in the way many if not most of the issues are addressed and resolved in practice in individual MS. The overview of European integrated hydro-economic models presented in the annex to this paper is based on the international workshop by the same name organized by the first author in November 2005 in the Institute for Environmental Studies, Vrije Universiteit Amsterdam (see www.ivm.falw.vu.nl/watereconomics).

The paper is organized as follows. Section 2 first outlines the economic elements in the WFD. This is followed in Section 3 by a presentation and discussion of the key issues surrounding the economic analysis in the WFD. Section 4 presents a description of the available economic methods and tools to address these issues, including cost-effectiveness analysis, cost-benefit analysis, non-market valuation of the economic benefits of water policy, multi criteria analysis, integrated river basin models, and economic instruments. A summary of the key issues and the available economic tools, methods and instruments and conclusions are provided in Section 5.

2. THE ECONOMIC ANALYSIS IN THE WFD

The WFD is one of the first European Directives in the domain of water, which explicitly recognizes the role of economics in reaching environmental water quality objectives. The Directive calls for the application of economic principles (e.g., polluter pays principle), approaches and tools (e.g., cost-effectiveness analysis) and for the consideration of economic instruments (e.g., water pricing methods) for achieving good water status for all European water bodies in the most effective manner. The Guidance Document on the Economic Analysis prepared in 2002 by the European Water and Economics Working Group (WATECO) advises that the various elements of the economic analysis should be integrated in the policy and management cycle in order to adequately aid decision-making when preparing the river basin management plans. The integration of economic reason and arguments throughout the WFD policy and decision-making cycle is presented in Figure 2.1.

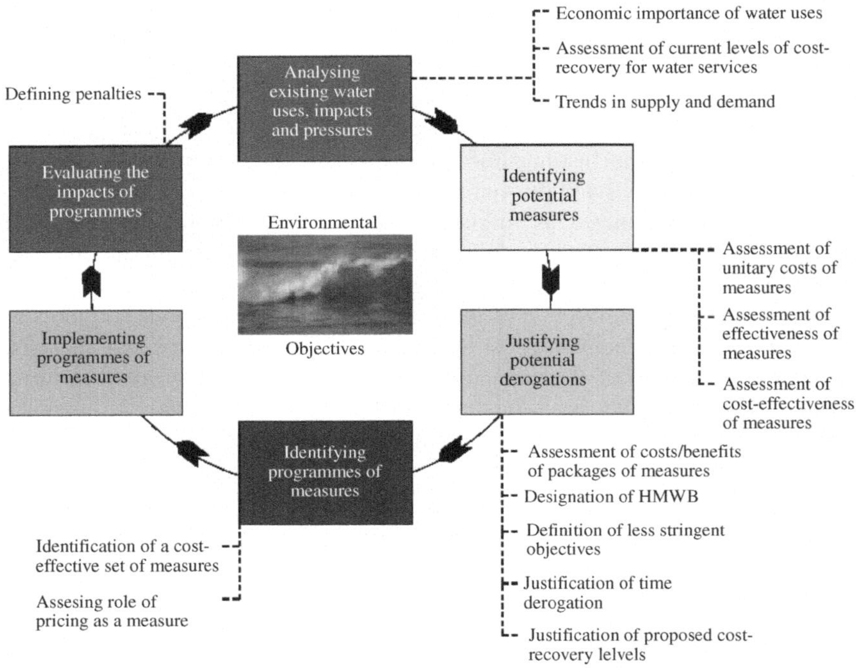

Figure 2.1 The role of economics throughout the WFD implementation process. (Source: WATECO Guidance Document)

The main elements of the economic analysis are found in Articles 5 and 9 and Annex III in the WFD. Economic arguments also play an important role in the political decision-making process surrounding the preparation of River Basin Management Plans in Article 4 where derogation can be supported by the strength of economic arguments when setting environmental objectives. The economic analysis can be summarised as follows:

(1) Economic characterisation of the river basin (Article 5)
  - Assessment of the economic significance of water use in the river basin.
  - Forecast of supply and demand of water in the river basin up to 2015.
  - Assessment of current cost recovery by estimating the volume, prices, investments and costs associated with water services in order to be

able to assess cost recovery of these water services, including environmental and resource costs.

(2) Cost-effectiveness analysis (Article 11 and Annex III)
  - Evaluation of the costs and effectiveness of the proposed programme of measures to reach the environmental objectives.

(3) Disproportional costs (Article 4)
  - Evaluation whether costs are disproportionate.

(4) Cost recovery and incentive pricing (Article 9)
  - Assessment of the distribution of costs and benefits and the potential impact on cost recovery and incentive pricing.

The main steps in the economic analysis identified by WATECO (2002) and the associated time path are illustrated in Figure 2.2. The first step, the economic characterisation of river basins, has been completed. In the next years the preliminary findings so far in the different European river basins regarding the main water users (sources of pollution) will be further elaborated (including a more detailed definition of environmental objectives) and a start has been made with the identification of additional measures needed to reach good water status in a second step.

By the end of 2007 each EU Member State had to produce an overview of its basic and additional measures according to Article 11, from which the most cost-effective programme of measures were to be selected in step 3 by the end of 2008. Based on a cost-effectiveness analysis of programmes of measures, the question whether the total costs of additional measures to reach good water status are disproportionate were to be addressed at the same time. Finally, the financial implications of the basic and additional measures for different groups in society had to be evaluated by 2009, including the level of cost recovery, the use of economic instruments (e.g., levies, taxes, water prices) and their role in achieving a more efficient and sustainable water use.

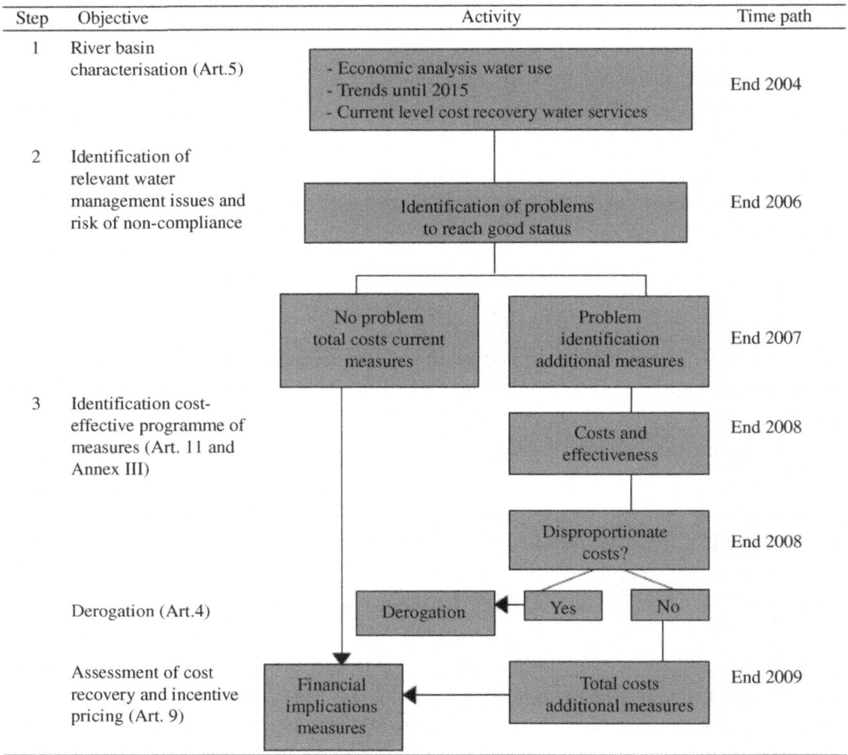

Figure 2.2 Steps in the economic analysis in the WFD and corresponding time path. (Source: modified from the WATECO Guidance Document)

3. KEY ISSUES IN THE ECONOMIC ANALYSIS IN THE WFD

This section discusses the key issues in the economic analysis underpinning the implementation of the WFD. These key issues are directly related to the different steps in the economic analysis presented in Figure 2.2., and include:

(1) Integrated water-economic river basin information systems and indicators.
(2) Selection of cost-effective programs of measures to reach the WFD objectives.

(3) Definition and assessment of disproportionate costs to reach the WFD objectives.
(4) Cost recovery of water services including environmental and resource costs.
(5) Economic instruments to reach the WFD objectives.

The first issue refers to the compilation of integrated policy (progress) indicators at river basin level and is relevant for the first two steps in Figure 2.2. Integrated progress indicators enable river basin characterization as required in step 1 from an environmental and economic system perspective and show the possible trade-offs involved (i.e., the core of economics), and the identification of possible water management problems as required in step 2. The following issues refer to step 3. The assessment of disproportionate costs follows the assessment of the least cost way to reach the WFD environmental objectives of good chemical and ecological status (cost-effective program of measures). This is followed subsequently by an assessment of the distribution of the financial burden across different groups in society (households, agriculture and industry) based on the principles of cost recovery and the polluter pays through existing or future market-based pricing policies as formulated in Article 9 in the WFD. The key issues will be elaborated in more detail in the sub-sections below.

3.1. Integrated water-economic river basin information systems and indicators

The implementation of the WFD has increased policy and decision-maker demand for integrated hydro-economic information at the level of river basins. The WFD requires that river basins across Europe are described in both physical and economic terms. According to Article 5 in the WFD, the economic characterisation of river basins should include an assessment of the economic significance of current water use and future water use up to 2015 (WATECO, 2002). Integrated hydro-economic accounting systems are considered to be useful tools, displaying information about the interactions between the physical water system and the economy at national and river basin scale. This involves not only linking water related environmental data to economic data, but also presenting these data at the relevant spatial and temporal scales through appropriate (dis)aggregation procedures. It is especially this issue of matching available data across various spatial scales, which has proven to be one of the major challenges in the compilation of integrated river basin information systems (Brouwer *et al.*, 2005).

Box 3.1 Example of integrated river basin accounts and policy progress indicators

The economic significance of water use in economic production processes in different river basins in the Netherlands is measured in the National Accounting Matrix including Water Accounts for River Basins (NAMWARiB) in two different ways. Economic significance is measured through production values and value added generated in river basins per sector (expressed in euros). Water use is measured through water extraction (expressed in cubic metres) by economic activities and the emission of polluting substances (expressed in kilograms) per sector. Water use can furthermore be measured through wastewater discharge per sector in each river basin (expressed in inhabitant equivalents). Based on time series analysis possible trends can be identified. Trends in economic driving forces are linked to pressures such as water consumption, wastewater production and the emission of polluting substances (nutrients, metals etc.). An example is given in the Figure 3.1 below. At the national level, real economic growth (in terms of GDP in constant prices) over the period 1996–2001 was 18 percent (on average 3 percent per year). Total wastewater production remained more or less the same over that same period, whereas the emission of nutrients decreased by approximately 15 percent and the emission of metals by about 10 percent. Hence economic activities seem to use the water environment in a more efficient way over this time period.

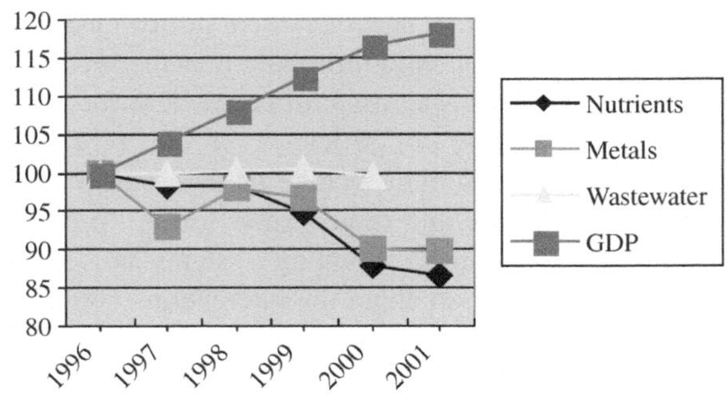

Figure 3.1 Economic growth, wastewater production, emission of nutrients and metals in the Netherlands (1996 = 100)

NAMWARiB also allows detailed trend analysis of specific substances per sector at river basin level. The trends identified at national level, are, for instance, also found in Rhine West, but in the Meuse river basin metals show a more fluctuating development, with an increase over the period 1998–2000 (Figure 3.2).

These diagrams have to be interpreted with the necessary care. They provide for instance no direct evidence of a linear link between production and water use such as wastewater production or the emission of pollutants. Nevertheless, the presented indicators are helpful in assessing the success (or failure) of environmental (sector) policy, as they provide important insight in the environmental efficiency of economic activities. They also provide a basis for trend analysis and the calibration and validation of integrated hydro-economic optimization and simulation models. Based upon the observed development of economic activities within sectors and corresponding water use over say the past 10 years, one can extrapolate this development into the future.

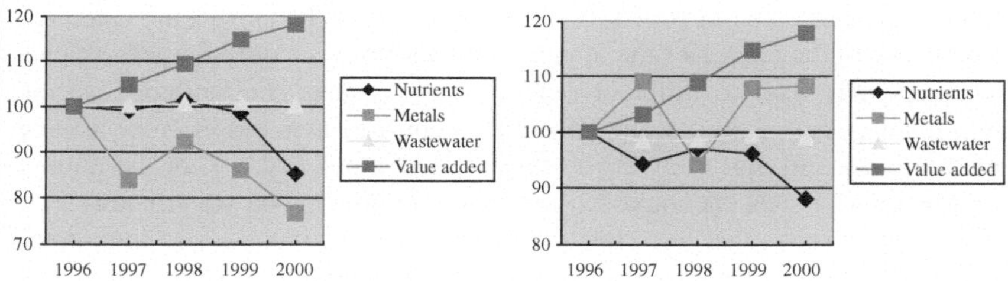

Figure 3.2 Economic growth, wastewater production,emission of nutrients and metals in the Rhine West (left) and Meuse (right) river basin over the period 1996–2000 (1996 = 100).
Source: Brouwer *et al.* (2005)

According to the European Task Force on Water Satellite Accounts (2002, p.4), the implications of the WFD include more focus on the geographical boundaries of available data, i.e., water bodies and river basin districts. Also the United Nation Handbook on Integrated Environmental and Economic Water Accounting acknowledges this (UN, 2006, p.17): '*... it is important that the spatial reference is the same for hydrological and economic data.*' However, '*river basins for which hydrological data is usually available, do not generally coincide with administrative regions for which economic information is collected*'. No consistent pan-European water economic information system exists, which allows for a consistent (dis)aggregation and comparison between the national and river basin level. Most efforts in the area of integrated water accounting are based on national statistics. An example of an integrated river basin information system, its use and usefulness, is presented in Box 3.1.

3.2. Selection of cost-effective programs of measures to reach the WFD objectives

Environmental efficiency has always been an important guiding principle in European environmental legislation, including the WFD. According to the WATECO guidance, the definition of the program of measures and the ranking of basic and supplementary measures based on cost-effectiveness criteria is the key economic input into the preparation of the RBMP. The main steps identified in the guidance include the estimation of the costs of each measure, the estimation of the effectiveness (environmental impact) of each measure and the ranking of cost-effectiveness of measures. Hence, the information needed basically relates to the costs of potential measures and their effectiveness (see the example presented in Box 3.2). According to WATECO, specific care needs to be given to the choice of the effectiveness indicator as different effectiveness indicators may lead to different rankings of measures. Furthermore, specific attention may be required as the effectiveness of measures can often be assessed quantitatively only for a few environmental indicators, and not for the range of environmental issues encompassed in the definition of good water status.

WATECO states that uncertainty about costs, effectiveness and time-lagged effects of measures needs to be dealt with throughout the economic analysis process, and more generally throughout the process of identifying measures and developing the river basin management plan. Sources for uncertainty are highly diverse according to situations and river basins, but will exist with regard to the assessment of pressures, impacts, baseline scenario, costs and effectiveness. The main sources of uncertainty underlying the selection of a cost-effective programme of measures include (Brouwer, 2005):

(1) Uncertainty about the environmental goals and parameters: i.e., the target situation and how the achievement of this target situation is measured. Both the outcome (good status) and probability of reaching the outcome are uncertain.
(2) Uncertainty about the sources of pollution: point sources or diffuse sources and the extent to which these sources (and pressures) contribute to (impact on) the water quality problem through the usually complex environmental source-effect chain in time and space. This includes uncertainty about possible future sources of pollution.

(3) Uncertainty about the effectiveness of proposed measures taken at source or the effect side of the ecology-economy interface on the ecology of the water system (again in time and space). Related to this is the uncertainty of reaching the imposed environmental standards.

(4) Uncertainty about the direct and indirect costs of proposed measures. Indirect costs are related to the fact that an economic activity has various forward and backward links to other economic activities. Interventions in one activity may therefore result in a chain reaction throughout the entire economic system, depending on the nature and the extent of the intervention. In order to be able to assess these multiplier effects accurately, a profound understanding of the structure of the economy at river basin scale is required (for instance based on integrated hydro-economic models; see the model description in Section 4), which is rather limited at the moment across EU Member States.

Based on experiences in various regional and international pilot river basins, (e.g., Boeters *et al.*, 2006; Trémolet Consulting, 2006) the following key issues arise when trying to identify a cost-effective program of measures:

Identification of the environmental objective(s) involved

The concrete environmental WFD objectives for different types of water bodies (artificial, heavily modified, natural) are still unknown. Especially when aiming to develop an international river basin management, including transboundary programs of measures, some degree of common understanding and consensus is needed about the environmental objectives involved, and the measurement of progress in reaching these objectives through compatible monitoring programs.

Box 3.2 Example selection cost-effective program of measures to improve water quality in the West Scheldt estuary

An example of a cost-effectiveness analysis is presented in the table below (adapted from Baan, 1996). The table presents possible measures to improve water quality in the West Scheldt estuary, located in the south-east of the Netherlands (red area in the map), their effects and costs. A measure's cost-effectiveness is calculated by simply dividing the annual cost (column 3) by the annual effect (column 2). The measure with the lowest cost per unit pressure reduction is the most cost-effective measure. In this particular study, dredging is the most cost-effective solution, followed by the treatment of drinking water and the dislocation of the effluent from the wastewater treatment plant (WWTP) outside the estuary. Replacement of zinc plates on ships is the most costly measure per unit pressure reduction. The pressure reduction in the table below is expressed in 'Pressure Equivalents' (PE). This PE is based on the measures' emission reduction of different pollutants (N, P, Cd, Cu, Zn) in kg per year (estimated through expert judgment) compared to the threshold concentration levels of each pollutant in kg/m^3.

Measure	Pressure reduction (PE in 10^6 m^3/year)	Cost (10^6 €/year)	Cost-effectiveness (€/10^3 m^3 PE)
Sanitation industrial discharge	1,435	2.0	1.4
Agricultural runoff	7,057	17.3	2.5
Treatment drinking water	663	0.3	0.5
Coating roof gutter	200	0.8	4.0
Additional wastewater treatment WWTP	3,276	5.5	1.7
WWTP effluent dislocation outside estuary	4,681	2.5	0.5
Replacement zinc containing paint for ships	1,467	1.0	0.7
Replacement zinc plates ships	13	0.1	7.7
Dredging	11,000	3.4	0.3
Reduction polder drainage	45	0.1	2.2

The Table above is illustrative of a number of important methodological issues when selecting a cost-effective program of measures. First of all, the objective (water quality improvement) and the effects of different measures on water quality can be measured through a variety of different physical indicators or parameters. An important question is how to make these indicators comparable and compatible across measures targeted at different pollution sources and how to account for the fact that some measures may impact on different water quality parameters at the same time. In this specific case, the effects of measures on different pollutants were expressed in a common pressure equivalent using a specific conversion procedure based on equivalent weights.

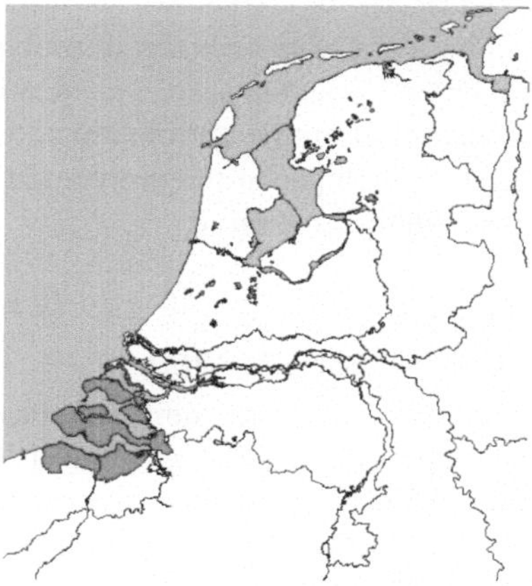

Figure 3.3 Location study area

Second, measures can be taken at different points along the environmental source-effect chain. Preventing instead of treating water pollution may be another important decision criterion for the selection of measures. Examples here are dredging versus the dislocation of the WWTP effluent discharge point. Third, given the spatial and temporal complexity of many water systems, pressure reduction in terms of the emission of kilograms of a pollutant per year is often more easily estimated and predicted than the actual impact of this pressure reduction on water quality in terms of milligrams of a pollutant per liter. Although it is water quality that is of prime interest in the selection and assessment procedure, pressure indicators are often used as proxies for the impacts on water quality given the uncertainties surrounding dose-response relationships and environmental damage functions.

Identification of the sources of pollution, pressures and impacts now and in the future over the appropriate time horizon

The exact relative contribution of different sources of pollution to the environmental water quality problem is often unknown, making it hard if not impossible to target different polluting sectors with specific measures to bridge the gap between the expected and desired state of the water body and river basin by 2015.

Quantification of the gap to be bridged

Related to the previous point, the gap to be bridged (problem definition) appears to be hard and in some cases even impossible to quantify given the lack of scientific knowledge and the uncertainties surrounding dose-effect relationships. In many cases a qualitative assessment of the problem at hand based on expert judgement is the best available estimation.

Identification of measures to bridge the gap between the reference (baseline) situation and target situation

The WFD distinguishes between basic and additional measures. Basic measures include measures, which help to achieve existing European water related legislation (e.g., Nitrates Directive or the Urban Wastewater Directive). Although the objectives of existing European and national policy are usually clear, it is less clear which types of measures have to be and will be implemented in the future in order to reach these objectives, not least because of the given uncertainties about their impact on water quality. Another important issue is the distinction between technical measures and economic instruments (see Section 3.6). In practice, the distinction is often not clear, and measures and instruments are used interchangeably in the cost-effectiveness analysis. Evaluating technical measures and instruments simultaneously has the risk of double counting. Often, instruments and measures are implemented sequentially, not simultaneously, and there is therefore a real risk of double counting.

Evaluation of the effectiveness of measures

The evaluation of the effectiveness of the proposed measures is probably the most troublesome in the entire decision process given the lack of knowledge and information and hence substantial uncertainties involved. Different indicators are used (e.g., chemical, biological), focusing on different points along the environmental dose-effect chain (e.g., pressure indicators and impact indicators), expressed in different units (e.g., kg/ha, mg/litre) making it hard to compare results from different types of measures in a compatible, meaningful way. Multi-criteria analysis (MCA) techniques (see the method description in Section 4) may be one possible way of dealing with this plurality of measurement units in a systematic and coherent way. Significant differences also exist between the temporal and spatial scale at which measures are implemented in a basin and the associated differences in temporal and spatial impacts.

Evaluation of the costs of measures

The distinction between different cost types (financial versus economic; fixed versus variable; direct versus indirect), cost items (e.g., one-time-off investment costs, annual management and maintenance costs, exploitation costs, overhead and depreciation costs) and their representation in time (e.g., annuity, (net) present value) is of utmost importance to arrive at a meaningful comparative analysis across measures and basins. It has to be made explicit which cost type and (proxy) estimation is being used, as there may exist significant differences between direct financial (engineering) costs and the usually broader defined direct and indirect economic costs. Different bottom-up and top-down calculation and estimation methods are available to assess the direct and indirect financial and economic costs of programs of measures (see the method description in Section 4). The way costs are estimated has important implications also for the assessment of disproportionate costs (see the next section).

Obviously, the ranking of measures in terms of increasing costs and based on this the selection of the least cost way to reach the environmental WFD objectives can only take place if the extent of the water quality problem (gap) is known and hence the necessary means (programs of measures) to solve the problem or bridge the gap, and the costs and effects of programs of measures are measured in comparable units. This requires inevitably the introduction of some degree of quantification in the analysis, including the qualification and quantification (in terms of orders of magnitude) of the main uncertainties involved.

3.3. Definition and assessment of disproportionate costs to reach the WFD objectives

The procedure to establish the environmental WFD objectives are described in Article 4. This article also allows for so-called derogation, i.e., to lower environmental objectives or delay them in time based on (i) technical feasibility of achieving the objectives and/or (ii) disproportionate costs (paragraph 3–7). The concept of disproportionate costs is not a standard economic concept. The assessment of disproportionate costs is subjective (WATECO, 2002), depends on the political economy of a country or river basin region, and proves to be surrounded by a lot of uncertainty as to their exact definition at what scale in the practical implementation of the WFD. It refers to a skewed distribution (discrepancy) between objectives and measures for reaching these objectives, but concrete national or international benchmarks do not exist, also not in comparable European Directives such as the IPPC Directive where the concept of Best Available Techniques Not Entailing Excessive Costs (BATNEEC) is introduced or the Habitats Directive, which refers to 'imperative reasons of overriding public interests, including those of a social or economic nature' to justify exemptions.

In theory, the assessment of disproportionate costs follows the assessment of the least cost way to reach the WFD environmental objectives and the corresponding distribution of the financial burden across different groups in society (households, agriculture and industry) through existing or future pricing policies as formulated in Article 9 in the WFD. However, in practice excessive or disproportionate costs often also play an important role already during the first stages of selecting a cost-effective program of measures (see previous section) as a political screening criterion[1]. Whether the costs of reaching the defined environmental WFD objectives are disproportionate depends on willingness and ability to pay by different socio-economic groups and sectors in society to whom these costs are transferred and who pay for them. Willingness to pay (WTP) depends on political, sectoral and public preferences for the environmental objectives involved and the environmental and resource costs of current water use (see Section 3.5) and ability to pay. Ability to pay in turn depends on the financial strength and capacity of the public and private actors and sectors involved, which is in part determined by the current state of the economy.

Given the definition of disproportionate: unequal relationship between two units (in this case the objective and the means to achieve the objective), the key question to assess disproportionate costs is to what extent the costs of WFD measures outweigh the benefits of reaching the WFD objectives. Cost-benefit analysis (CBA) would be the appropriate economic tool to assess this relationship (see the method description in Section 4), using the net present value or benefit-cost ratio as the decision criterion, comparing all positive and negative welfare effects of the WFD measures to achieve good chemical and ecological status. Assuming that all positive and negative welfare effects are accounted for in the CBA, a negative NPV or a benefit-cost ratio less than one, indicating a national welfare loss, would result in principle in a rejection of the proposed program of measures to reach the proposed WFD objectives. However, given the public good nature of the WFD objectives and in the interest of future generations, the government may be willing to accept a benefit-cost ratio of less than one. The

[1]The practical selection procedure of a cost-effective program of measures may in this sense be better represented in terms of a multi-criteria analysis (see the method description in Section 4).

question then becomes how many times the costs can exceed the benefits before the program of measures involved is considered disproportionate. Uncertainty is expected to play an important role here too, related to the assessment of the costs and the environmental and social benefits of WFD implementation.

Another important issue is the scale at which the assessment of disproportionate costs takes place. Costs may be considered disproportionate at the level of an individual region or sector given its financial buoyancy, but economically beneficial at the higher aggregated level of a nation as a whole. On the other hand, the skewed distribution of costs and benefits across time and space and across different sectors may be an important reason for policy makers to consider a particular solution undesirable given for example the disproportionate effects on certain vulnerable groups in society. It is important to realize that the answer to the question whether the estimated costs of programs of measures are disproportionate remains subjective in all these cases (see also Box 3.3).

Box 3.3 Example of disproportionate costs in agriculture

In 2003 the Dutch Ministry of Agriculture carried out a quick scan of the possible implications of the implementation of the WFD for the agricultural sector. Agriculture has a relatively large share in the emissions of nutrients and pesticides to water. Compared to a 2015 baseline scenario, the consequences of two possible WFD policy scenarios were assessed in terms of reduction of area size, net value added and employment. Policy scenario A corresponds to a modest WFD policy objective, while policy scenario B refers to a high ecological ambition scenario (see Table below).

Scenario	Area size (%)	Value added/ha (%)	Total value added (%)	Employment direct (%)	Employment Indirect (%)
Baseline	100	100	100	100	100
Scenario A	35	94	33	36	36
Scenario B	0	0	0	0	0

Source: van der Bolt *et al.* (2003)

The study shows that in the case of scenario A agricultural land is reduced to a third of the expected area size under the baseline scenario and that in the case of scenario B agriculture disappears completely. The decrease in value added per hectare is relatively low, but the decrease in total value added is substantial due to the reduction in area size. In some areas, especially peat clay areas, the emission reduction objectives are not met under scenario A, also not when taking all agricultural land out of production, mainly because of the long time it takes for all accumulated nutrients to run off from the phosphate saturated soils. In the case of scenario A, the environmental objectives are not met anywhere in the Netherlands and the question therefore is how realistic and feasible the chosen policy scenarios are.

In an official reaction, the national farmer association (LTO) claims that the implications of the WFD for the agricultural are '*completely disproportional*'. Nature conservation organizations, on the other hand, use the study results to show that the agricultural sector is not sustainable when it comes to the protection of water quality and ask the Government to promote a more sustainable water use by the sector.

Source: Brouwer *et al.* (2005).

3.4. Cost recovery of water services

According to Paragraph 1 in Article 9 in the WFD, '*member states shall take account of the principle of recovery of the costs of water services, including environmental and resource costs, having regard to the economic analysis conducted according to Annex III, and in accordance in particular with the polluter pays principle*'. The assessment of environmental and resource costs will be addressed in the next section. This section focuses specifically on the economic cost assessment of water services. Water services are defined in paragraph 38, article 2 in the WFD: 'Water services means all services which provide, for households, public institutions or any economic activity: (a) abstraction, impoundment, storage, treatment and distribution of surface water or groundwater; (b) waste water collection and treatment facilities which subsequently discharge into surface water.'

The WATECO guidance identifies four different types of information necessary to assess the current levels of cost recovery:

- Estimation of the costs of water services, including financial, environmental and resource costs. The financial costs broken down into operating, maintenance and capital costs.
- Estimation of the price or tariff currently paid by the users and evaluation of tax transfers.

- Assessment of the extent of financial and environmental cost recovery by water service and sector.
- Assessment of the contribution to cost recovery from key water uses.

If felt necessary, a review of incentive pricing properties of existing tariffs can be initiated on the basis of the information above. Principles for allocating costs of water services to categories of water users will need to be defined in a coherent manner. The main output of this element in the economic analysis is an assessment of the current extent of cost-recovery. Assessing incentive pricing properties of existing tariffs may be difficult in practice, but is an essential step so as to inform the future introduction of incentives in tariffs by 2009.

In practice, the distinction between water use and water services is not very clear-cut and surrounded by many political obstacles related to the discrepancy between current and potentially future pricing of water use and services. The WATECO guidance regards water services as the link between the environment and water use, consisting of activities, which change important characteristics of water or wastewater. It is this change in water or wastewater characteristics, which the WATECO guidance calls a water service. Water use is defined as an activity, which has a significant impact on a water body's status. It is unclear, however, what is considered a '*significant*' impact.

The issue of water use and water services can also be addressed from a different perspective, namely that of the water system instead of the human activities related to the water system. This approach is in line with the ecological principles underlying the WFD, and is based upon the international environmental and ecological economics literature, in which it is nowadays generally acknowledged that ecosystems, including aquatic ecosystems, provide goods and services (functions), which are beneficial to human society (e.g., Adamus and Stockwell, 1983; Nichols, 1983; Larson *et al.,* 1989; Turner, 1993; Barbier, 1994; Gren *et al.,* 1994; Maltby *et al.,* 1995; Costanza *et al.,* 1997). Following this approach, a distinction can be made between the supply of and demand for water services. This introduces basic economics back into the economic analysis of the WFD, where demand for water services is reflected in water use.

Water services are supplied first of all by the water system itself. Examples are basic services such as surface or groundwater used for drinking water purposes or growing crops, but also the nutrient and wastewater absorption capacity of aquatic ecosystems are important services provided by the water system. In the environmental economics literature the former are called direct functions and the latter indirect ecosystem functions. Human activities may also add value to these services or try to (partly) replace these services, but the economic significance of these activities usually depends largely on the original services provided by the water system. Examples are pumping up surface or groundwater, transportation of water, water purification, water distribution etc. In efficiently organised modern societies, these activities are nowadays considered essential services in themselves, satisfying consumer demand for water availability and water quality. Hence, depending on demand, water services may be modified to different degrees by human activities in order to make them suitable for final consumption. Examples of human activities, which (partly) replace original water services, are wastewater collection and treatment. Often, the original waste absorption capacities of aquatic ecosystems have been impaired by excess pollution and hydro-morphological modification. The original water services or the water services modified by human activities are supplied to various groups of consumers, such as households, agriculture and industry. In some cases these consumers are also the suppliers of the water service. This is labelled 'self service' in the WATECO guidance. Examples are farmers or industries, who pump up surface or ground water for their own use such as crop irrigation, cooling water, paper production or food processing.

The principle of cost recovery implies that the economic production costs of a good or service – measured through their opportunity costs - are fully recovered. An example is given in Box 3.4. The cost of a productive activity consists, in theory, of the opportunity costs of the necessary inputs. The opportunity cost of employing an input is the highest net benefit generated had it been employed elsewhere. If input and output markets exist and function well, the opportunity costs are reflected in the market prices paid for the inputs. For example, the price of the necessary labour, equipment or electricity to produce a good or service.

However, in some cases, inputs are used or outputs produced for which no markets exist and hence also no market prices are available to reflect their opportunity costs or which are traded in distorted markets, which do not reflect the real opportunity costs of the inputs used or outputs produced. These inputs or outputs usually involve so-called public goods or services, i.e., goods or services (consumed or produced), which can technically (and often also institutionally or politically) not be broken down in separate and individually purchasable units. Examples include resources such as air and water and their natural purification processes (or their function as a disposal sink for emitted pollutants). Here, environmental damages caused by water use (e.g., water abstraction or the emission of pollutants) often do not result in a private cost to individual agents, but do result in a social cost to society (Brouwer *et al.*, 2004). This is what the WFD refers to as environmental and resource costs (see the next section).

Box 3.4 Example of the price of water and cost recovery

In the Netherlands, most water management costs are recovered by charging the users of the services provided. Water pricing policy is based on the beneficiary and polluter pays principle. Still, pricing and financing structures can be complex. For example, consumers of drinking water pay drinking water companies the production costs of drinking water, and additionally also indirectly the general value added tax (VAT), the State groundwater tax, the Provincial groundwater levy and a tax on tap water (Figure 3.4). The groundwater tax and levy can be considered as a way to internalize the external social costs of groundwater extraction and the associated environmental and resource costs, even though the groundwater tax is not earmarked, meaning that it is not actually used to compensate for any environmental damages caused by groundwater extraction.

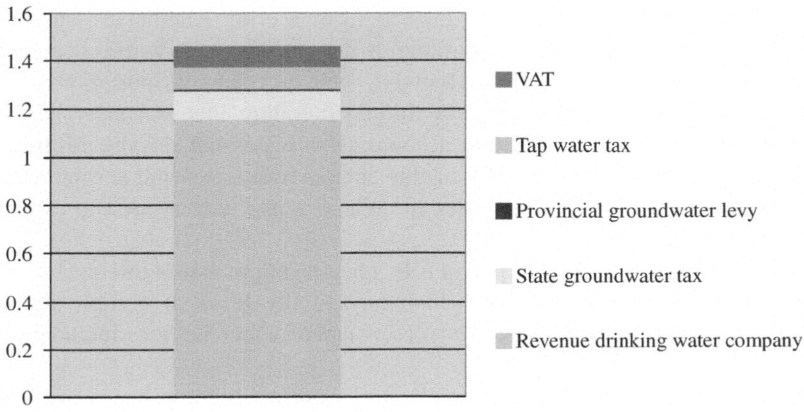

Figure 3.4 Average drinking water price and its components in €/m^3 in 2001 in the Netherlands
Source: VEWIN (2002)

Besides environmental damage not accounted for through market prices (i.e., market prices only reflecting the private opportunity costs and not the social opportunity costs), Renzetti and Kushner (2004) distinguish a number of other reasons why a government agency or firm's private accounting of the costs of its activities may differ from the full economic cost of those activities. First, there may be accounting guidelines set out in government regulations, which dictate the way in which costs are recorded. For example, utilities may be prevented from assigning a competitive rate of return as part of the opportunity cost of their purchased capital goods.

Secondly, a water utility may receive subsidies from other agencies. These could include direct subsidies such as capital grants from higher levels of government or indirect subsidies as might occur if a municipal water utility were to receive services from the city's legal department without charge.

Thirdly, analysts and utility regulators may disagree in how completely they wish to see an agency move to full cost accounting. For example, some may argue that it is sufficient to see that operation and maintenance costs are fully accounted for. Others may argue that operating and maintenance plus capital costs must be fully accounted. Still others would argue that also external costs such as environmental damages and the opportunity cost of raw water supplies must be included.

A fourth factor that inhibits implementation of full cost recovery is a lack of standardized guidelines for these types of accounting. Although there exists a large body of economic research devoted to measuring whether a household, government agency or firm's accounting of the costs of its activities accurately reflect the costs borne by all of society, such as road transportation or the generation of electricity[2], there is a limited literature that is concerned with applying the principles of full cost accounting to water and sewage utilities.

Finally, because of the novelty of this approach to water and sewage utility accounting, there may be difficulties in collecting the data needed to estimate some cost components.

Other important issues that play a role when trying to assess the level of cost recovery of water services and which add to the level of complexity and uncertainty when trying to assess cost recovery of water services include:

- In some cases there does not exist a clear direct relationship between the service in question, the provider of the service and the user of the service, making it hard if not impossible to assess to what extent different water users contribute adequately to cost recovery as required in article 9. For example, the management of rivers and watercourses (e.g., dredging)

[2]In Europe, a major EU funded research study was undertaken over the past 10 years to estimate the external costs to the environment and human health of electricity production from fossil, nuclear and renewable sources (ExternE: Externalities of Energy). A recent application of these ExternE cost indicators is the CBA of the Clean Air for Europe (CAFE) program (Holland *et al.*, 2005).

serves both commercial and recreational shipping and flood control. Flood control management is another example of the provision of a public good, which cannot technically be broken down in individual units and sold to individual beneficiaries.

- In other cases there exist interdependencies between water services and their costs. For example, an increase in the level of sewerage provision levels (e.g., increase in connection rate) will also result in an increase of water treatment costs.
- Not all costs are always clearly visible. For example, environmental and resource costs (see the next section). Or costs are financed through different financial channels and by different institutions (e.g., local tax revenues and general income taxation). This too adds to the problem of assessing the adequate contribution of different water users.

3.5. Environmental and resource costs

Environmental and resource costs are mentioned in article 9 related to cost recovery (see Section 3.4), but are expected to also play an important role in the assessment of disproportionate costs in terms of the public benefits foregone of not reaching a good chemical and ecological status. In the Wateco guidance's glossary of terms, **environmental costs** are defined as representing *the costs of damage that water uses impose on the environment and ecosystems and those who use the environment (e.g., a reduction in the ecological quality of aquatic ecosystems or the salinisation and degradation of productive soils).* **Resource costs** are defined as *the costs of foregone opportunities which other uses suffer due to the depletion of the resource beyond its natural rate of recharge or recovery (e.g., linked to the over-abstraction of groundwater).*

However, the distinction in the Wateco guidance between environmental and resource costs is not clear-cut. Also their estimation and use are still surrounded by a lot of discussion. It is not so much the general idea that water use imposes a cost on the water environment and those who use this environment, which often remains unaccounted for in policy and decision-making, that is disputed. Cause for debate is primarily (1) the explicit distinction between environmental costs and resource costs, (2) their practical measurement, either through cost or benefit based approaches, and (3) the extent to which these costs have already been accounted for in existing pricing and financing mechanisms.

Although no explicit distinction is made in the existing environmental economics literature between environmental *and* resource costs (for an overview see Schaafsma and Brouwer, 2006), the terms have been interpreted and defined as two separate concepts in the Wateco guidance and subsequently also in the

ECO2 information sheet[3]. The distinction between environmental and resource costs in the Wateco guidance seems to be based primarily on the question whether the water resource is depletable (non-renewable) or not, while the ECO2 information sheet introduces the notion of use and nonuse values into the equation, where environmental damage costs could be argued to refer to nonuse values primarily attached to a healthy functioning aquatic ecosystem and the opportunity costs of those who use the water environment to the corresponding use values. Both the Wateco guidance and the ECO2 information sheet use the term 'opportunity costs' to describe and delineate resource costs and this seems to be considered a distinctive feature between environmental and resource costs. However, the concept of opportunity costs is a generally applicable concept in environmental economics, and may equally apply to environmental and resource costs.[4]

An important reason for their distinction is also to account for the fact that across Europe important differences exist in terms of water quantity and water quality management issues. The distinction is largely driven by concerns in South European MS facing (extreme) water scarcity and allocation problems, nowadays in the context of the WFD more generally referred to as resource costs. Resource costs here refer to competing and/or conflicting water use for different socio-economic activities (domestic household, agriculture, industry), primarily as a result of its limited quantitative availability in time and space (see also Section 4.7), but not exclusively as there exists a close relationship between water quantity (scarcity) and water quality. Water quality issues (pollution, eutrophication, loss of biodiversity) generally are the most important water management concern in North European MS, usually referred to as environmental costs.

More important than the discussion about the exact definition and distinction between the two terms is the question (1) what exactly constitutes damage, to the water environment and those who use the water environment due to excess pollution and/or abstraction, (2) what are the benefits of WFD implementation

---

[3]ECO2 was a drafting group established in 2003 under Working Group B 'Integrated River Basin Management' in the WFD Common Implementation Strategy (CIS) to address and advise the Working Group on the specific issue of environmental and resource costs, resulting in an information sheet.

[4]Water resources provide important commodity and environmental benefits to society. As such, the management of water is an economic, social and political issue encompassing all sectors of an economy, and involving trade-offs between competing users, as well as between additional economic growth and aquatic ecosystem protection and further water resource depletion and degradation. Any particular use of water will be associated with opportunity costs, which are the benefits foregone from possible alternative uses of the resource.

aimed at more sustainable water use (reducing or eliminating excess pollution and/or abstraction), and (3) how are these damage costs and WFD benefits best measured (in a valid and reliable way).

In theory, damage arises when there is a discrepancy between some reference and target point or situation. The latter can be measured, for instance, through existing environmental norms or standards or the right people attach to a clean environment and the provision of sufficient and clean water. In practice, sometimes also a point in the past, when pollution levels and corresponding damage costs were lower, is taken to represent the target situation. An example is the discharge of waste water into a water course at a rate (e.g., tons of N per year), which exceeds some permitted rate (in tons per year) and hence results in a eutrophic water system with negative consequences for both the biological diversity of the water system and the recreational amenities provided by the water course, including the possible negative effects on human health if the specific water body is also used for recreational swimming. In the context of the WFD, it seems logic to use the expected water status in 2015 as the reference situation, but other reference situations may also be appropriate. The same applies to the target situation. It seems logic to relate the target situation to the environmental objectives of the WFD, i.e., good ecological water status in 2015. However, other target situations may also apply. Hence, there is a strong dependence of environmental and resource costs on the physical status of the water system and knowledge and information about this physical status. This includes the damage caused to the water system as a result of pressures exerted on the water system, such as the extent to which the system's natural rate of recharge or recovery has been impaired by a specific water use. Other general damage categories include eutrophication, salinisation, dessication, loss of biological diversity and morphological changes to a water system. This type of information is crucial to the subsequent economic estimation of environmental costs. The physical status of a water body or water system provides the basis for the estimation of the environmental and resource costs in economic terms. If this information is not available, environmental and resource costs cannot be assessed.

Coming to the economic valuation of the costs and benefits of water use, a first important observation is that such economic valuation studies associated with the specific scope and objectives of the WFD are rare or non-existent in Europe and elsewhere. An example is given in Box 3.5. The main problem when considering economic choices related to water is that a competitive, freely functioning market does not exist for many water related uses. Many services provided by water resources are cost free, and users are not charged for their use (see Section 3.4). The reasons for this stem from water characteristics such as the fact that (a) water is an essential commodity the value of which for a basic survival amount is infinite, (b) water has natural monopoly characteristics, (c) property rights for water resources are often absent and difficult to define, (d) water is a 'bulky' commodity, thereby restricting the development of markets beyond the local area (see the next section about market based instruments).

Given the absence of a functioning market mechanism for many water uses and services, and in line with the increasing water conflicts and need for more efficient allocation, it is necessary to have knowledge and information of the marginal value or benefits of the resource in its alternative uses. The main aim of economic valuation is to provide the value information that would normally be provided in the market for efficient water resource allocation, including pollution (see the method description in Section 4).

Box 3.5 Example of environmental and resource cost types in the UK

England and Wales have a long tradition in the assessment of environmental and resource costs and will apply and develop this knowledge base to help implement the WFD. Previous research shows that these costs are highly significant, although their exact estimation is subject to a high degree of uncertainty on both scientific and economic grounds. Current information on the level of these damages is set out in the Regulatory Impact Assessment prepared for the WFD. These costs need to be properly assessed and quantified in order to develop the right set of measures for the WFD. A distinction is made between external environmental and financial costs and internal financial costs:

1. External environmental costs

External environmental costs are the damage costs (or loss of welfare) from current abstraction and discharges (after current controls).

2. Internal financial costs of current control measures

Considerable financial costs have already been incurred to control discharges, releases and abstractions affecting the water environment. For instance, water companies in England and Wales in the years 2000 to 2005 invested between £4.5 to £5.5 billion in order to address environmental impacts related to the discharge of pollutants and water abstractions. Similar capital costs were incurred in each of the two prior 5-year periods since the 1989 privatization of the water companies. Other bodies also bear financial costs in meeting existing standards.

3. External financial costs of control/abatement measures

The water companies' expenditures on water treatment include about £313million per annum on removing nitrates and pesticides and reducing risks associated with *Cryptosporidium* along with a number of other parameters. This represents about 10% of the total public water supply costs in England and Wales. About £240m of these costs are attributable to external sources such as agriculture. These represent external financial costs incurred by the water companies to treat pollutants originating from other sectors, most notably agriculture and other diffuse sources of pollutants, who do not pay for these costs.

£m 2002–03 prices		Annual costs bome by water company customers	% Contribution due to external sources	Total annual remediation costs attribute to external sources
Capital costs	*Nitrates*	47.7	80%	38.20
	Pesticides	78.3	89%	69.69
	Other parameters	108.6	50%	54.32
	Cryptosproridium	23.3	90%	20.97
	Total	234.7		162.2
Opex	*Deteriorating raw water quality*	70.4	69%	71.1
	Cryptosproridium	8.3	90%	7.5
	Total	78.7		78.6
Total		313.4		240.8

Source: ERM (2004) and Pretty (2000).

Disentangling these costs per type of pollutant and per source sector and identifying them is a first step to establish the link between pressures and impacts, which will help to devise more coherent policies. However, care should be taken as these three distinctly different types of costs above cannot be added since they are not comparable.

So, in summary:

- Cost 2 (already internalized financial control costs) represents the costs already incurred recently or in the past on treatment and control measures to reduce discharges or abstractions to their current levels.

- Cost 3 (external financial control costs) represents the portion of these water treatment costs paid by water companies that are to treat pollutants coming from other sectors (e.g., agriculture) who are not paying for these costs.
- Cost 1 (environmental damage costs) are the external residual environmental damage costs arising from current discharges, releases and abstractions (after the current controls which are included in the financial cost in 2 above).

Finally, it is important to point out the difference between price and value. Price and value are two distinct concepts. Water *per se* or water of a certain quality has value, but may have no price. Alternatively, water (of a certain quality) may have a price, but this price does not reflect the true economic value of water (or the value of a specific quality level). Hence the need to calculate and impute shadow prices for water use and the goods and services provided by aquatic ecosystems under circumstances where existing prices are believed to undervalue or overvalue the resource and economic efficiency is an important guiding principle in water (allocation) management. For the estimation of shadow prices different methods are available (see the method description in Section 4). These estimated shadow prices can subsequently be used to calculate the total economic value of a specific water use (e.g., the amount of water consumed times its shadow price), including environmental and resource costs of water use. Furthermore, existing pricing systems can be modified based on these shadow prices to allow economic market mechanisms of supply and demand re-allocate scarce water resources across different water uses and users and reduce possible inefficiency in current resource use (see the next section about economic instruments).

3.6. Design, application and evaluation of economic instruments

The WFD requires in article 9 that water pricing policies by the year 2010 provide adequate incentives for users to use water resources efficiently and thereby contribute to the WFD environmental objectives. The WFD furthermore refers to the use of economic instruments as part of the program of measures in its consideration 38 and in article 11. However, the role of economic instruments in the programs of measures is unclear, and their costs of implementation highly uncertain (Brouwer, 2005) (see Section 4.8). In many cases they are included as separate 'measures' in the programs of measures. In practice, economic instruments and technical measures are often complementary and implemented consecutively.

Currently, many water users do not pay for many if not most types of water use in many EU MS. The most important pricing mechanisms currently in use in EU MS are market prices for drinking water and taxes and charges for wastewater collection and wastewater treatment paid by domestic households and industry. Often, the charges and taxes are set at levels, which just cover direct financial costs, but lack incentives for more efficient and sustainable water use. Domestic household and industrial water demand are generally characterized by low price elasticities (e.g., Dalhuisen *et al.*, 2003), the same applies for irrigation water demand in agriculture (OECD, 1998; 1999). Additionally, in a number of EU MS agriculture and industry also pay an emission or pollution tax, charge or levy for direct discharges of pollutants to surface water and/or a groundwater or surface water abstraction charge, tax or levy. For all other kinds of water use economic instruments for efficient water use are usually absent.

Economic instruments, often also referred to as market-based instruments, are used to correct for the failure of markets to protect the (water) environment and face increasing popularity (e.g., EEA, 2005; 2006). The most important economic instruments are emission charges and environmental taxes where economic agents pay for their use of scarce environmental assets, and emission trading where emission quota are allocated and traded so that a unit price results, which reflects the environmental scarcity. Although the latter type of instrument has distinct economic advantages over 'command and control' (e.g., pollution source based standards) and taxes and charges, its application is very limited across Europe. Most experiences with tradable emission permits and water quality are found in the US (see Box 3.6).

Box 3.6 Experiences with tradable emission permits in the United States

Although the 1972 US Clean Water Act (CWA) does not foresee in any emission trading system (contrary to the Clean Air Act), various experiments have been carried out since the 1980s with tradable emission permits. In 1996 the US Environmental Protection Agency (EPA) published its 'Draft Framework for Watershed-Based Trading' and five demonstration projects were funded by the EPA and the Water Environment Research Foundation. The official EPA 'Water Quality Trading Policy' was established later in 2003, supporting the trade in nutrients (phosphorous and nitrogen) and sediments. Water quality trading can take place for impaired and unimpaired waters, that is, water bodies that do and do not meet the water quality standards. In the case of unimpaired water bodies, the trade is set up to compensate for new or increasing discharges. In the case of impaired water bodies, one of the following two conditions has to apply:

(i) Anticipating the establishment of the Total Maximum Daily Load (TMDL) for the water body involved, the emission trade reduces the degree of pollution and water quality moves towards the expected water quality standards.
(ii) After the TMDL has been fixed, the emission trade reduces the costs of the required emission reduction.

The US EPA itself does not organize the emission trade, but stimulates the authorities responsible for water quality management to develop and implement water quality trading programs based on the instrument's effectiveness and efficiency in reaching the water quality objectives.

General characteristics of the water quality trade in the US are:

- *Geographical area*: The trade takes place within a hydrological unit (watershed) or another area for which a TMDL has been established and approved. Usually a lake or part of a watershed.
- *Institutional embedding*: Pollution reduction 'credits' can be obtained if the reduction of polluting substances goes beyond the required TMDL. Consistency with the CWA has to be guaranteed and the emission trade can be incorporated in the National Pollutant Discharge Elimination System (NPDES).
- *Validity of the permits*: The time period is variable (month, season, year), depending on the specific requirements in the NPDES permit.
- *Parties involved*: The trade is bilateral: there is a direct relationship between the pollution source that generates the 'credits' by reducing its pollution or emission level and the pollution source that buys the 'credits' in order to be able to discharge its excess pollution in the water system. This is an essential difference with, for example, the European trading system of greenhouse gas emissions, where permits are traded on a market anonymously. Hence, product, price and trade conditions differ per transaction, limiting competition.
- *Pollutants*: Trade occurs in principle for nutrients and sediments only. Trade in other substances is possible (the EPA considers pilot projects for persistent bio-accumulating toxic substances), but only after thorough examination to ensure compatibility with existing water quality standards. Cross-pollutant trading is also possible, but again only if sufficient

information is available about the contribution of the substances involved to the water quality problem.

- *Non-point source pollution*: The extent of the pollution from non-point sources is hard to establish with complete certainty. Therefore, when trading with non-point pollution sources often non-point to user trading ratios are used of more than 1:1, implying that the non-point source has to reduce more than the quantity the point source is allowed to discharge based on the obtained 'credit'.
- *Monitoring and control*: The states and water managing authorities within a state are responsible for developing procedures to monitor and evaluate whether all conditions have been met and the pollution source that generates the 'credits' has actually reduced its pollution levels. If not, the permit keeper who bought the 'credits' still has to comply with the imposed TMDL as before. To overcome this problem, some states, like Michigan, have established additional rules for parties that acted in good faith, especially in the case of non-point source polluters. In the case of point sources with a NPDES permit, failure to meet the permit requirements is regarded as a violation of the CWA and penalties and sanctions are imposed accordingly to the CWA.

Factors and conditions that helped making the introduced emission trading system successful, include (a) mechanisms that facilitate the search for suppliers and users of pollution 'credits' and provide sufficient information to increase the market's transparency and reduce transaction costs (see for example the World Resources Institute website www.nutrientnet.org), (b) clarity and legal certainty about the value of the 'credits', institutionally embedded in the 2003 US EPA 'Water Quality Trading Policy', and (c) the absence of a variety of rules and regulations (such as the IPPC, Urban Waste Water Treatment and Nitrates Directive in the EU), which allowed especially diffuse sources to earn 'credits' by allowing them to find the best environmental practice for their specific activity given the specific conditions under which the activity is carried out.

Source: Adapted from Oosterhuis (2006).

Economic instruments aim to change human behaviour by financially penalizing undesired human behaviour or economic activities (e.g., water pollution) and financially rewarding desired behaviour or activities (e.g., water savings). In a market economy goods and services are allocated through the price mechanism. Market prices reflect demand and supply conditions for the goods and services involved. In a well-functioning market, these market prices include public preferences for environmental quality and the environmental and resource costs associated with water resource use. An important condition for such a well-functioning market is that the goods and services can technically be broken down in exclusive units with tradable entitlements and rivalry consumption. That is, consuming one unit (e.g., one m^3 water or emitting one kg N into surface water) excludes and is at the expense of consumption of that same unit by someone else. Water resources usually do not meet these requirements, the goods and services provided by these resources can be characterized as public goods. Non-rival and non-excludable goods and services like water-based recreation (in the absence of any congestion effects) or recycled water cannot be traded like 'normal' market goods and services, based on market prices that are determined through supply and demand. Water resource use is usually non-exclusive (in principle everybody has access to it) and non-rival (in principle consuming one m^3 water or emitting one kg N into the water does not affect the value of the entire flow of goods and services supplied by the water resource), implying that it cannot easily be financed by market user fees or beneficiary charges (Randall, 1983). See also the previous section.

Article 10 in the WFD requires MS to ensure the implementation of emission control based on best available techniques (BAT), emission limit values or, in the case of diffuse pollution, best environmental practices, cross-referencing to other EU legislation such as the IPPC Directive (96/61), the UWWT Directive (91/271) and the Nitrates Directive (91/676). The combined approach of pollution source based and water quality based requirements seems to leave little opportunity for economic instruments. Each and every source has to comply with stringent BAT or comparable standards. Allowing less efficient pollution sources to buy pollution credits for example from more efficient pollution sources and hence to exceed the imposed standards, is only possible if the WFD and other relevant EU laws were to be changed. This is what happened in the case of the EU greenhouse gas emissions trading system (Directive 2003/87). The IPPC Directive was amended so as to exclude emission limit values for greenhouse gases in permits for installations participating in emissions trading. Similar changes for water quality trading in an emission permits market will only be feasible if such trading is introduced on an EU-wide scale given the transboundary nature of many large rivers. Such an emission trading system is expected to be limited to fill the 'gap' between the water quality level that can be achieved with the currently available BAT to reduce pollution levels at source and the WFD required water quality level. This seems to reduce the scope for the introduction of a market based trading scheme, but filling the remaining 'gap' with market based instruments such as an emission permits system may be highly relevant in view of the fact that the 'gap' is found on the right-hand side of the cost-effectiveness curve (see method description in Section 4), where the marginal cost per unit pollution (emission) reduction is relatively high.

4. ECONOMIC METHODS, MODELS AND INSTRUMENTS

4.1. Introduction

In this section, we will present an overview of the available economic methods, models and instruments to address and deal with the key issues identified and described in the previous section. Figure 4.1. once again shows the different steps in the economic analysis in the WFD (see Section 2) and relates the steps and the activities under these steps to the available methods, models and instruments.

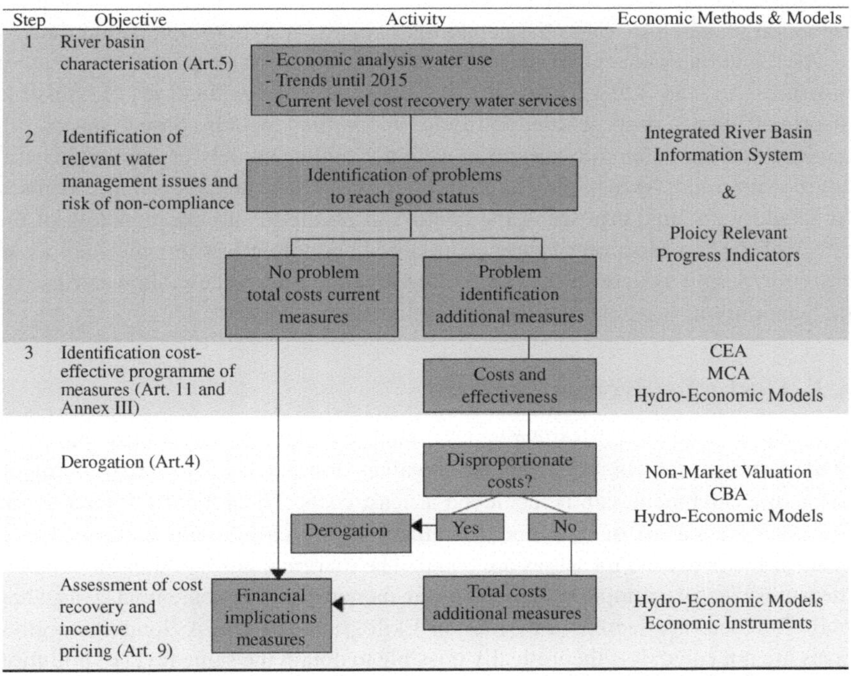

Figure 4.1 The role of economic methods, models and instruments in the economic analysis in the WFD

The main methods include cost-effectiveness analysis (CEA), cost-benefit analysis (CBA) and multi-criteria analysis (MCA). It goes without saying that CEA underpins the selection of a cost-effective program of measures in step 3. However, if these measures are screened and selected on the basis of other additional criteria too, as we saw in the previous section they may be, MCA may be a more appropriate method. Depending on the scale of WFD implementation and the required level of detail regarding the economic impacts of the proposed programs of measures, hydro-economic models can also be used to estimate the full economic costs and effects of the measures.

In the previous section the use and usefulness of CBA was already briefly discussed in the context of disproportionate costs. Non-market valuation methods provide important additional information for a fully monetized CBA given the public good nature of the expected benefits derived from the WFD implementation. One specific valuation method will be addressed in more detail here and that is benefits transfer. Benefits transfer is often used in policy analysis and appraisal. It is an attractive low cost alternative for carrying out original valuation research to value the economic benefits of WFD implementation.

As for the assessment of the full costs and effects of proposed programs of measures in the CEA, also in the case of the evaluation of possible disproportionate costs extended hydro-economic models based on a full coverage of the economic system (e.g., input-output models or applied general equilibrium models) may play an important role. These models can also be used for shadow pricing purposes in the context of article 9, and the modeling of the effect of the introduction of new or modified economic instruments such as an emission permit system to stimulate more efficient and sustainable water use on the economy as a whole and the various economic sectors.

4.2. Cost-effectiveness analysis

The purpose of a cost-effectiveness analysis (CEA) is to find out how predetermined targets, e.g., threshold values for nutrients or other pollutant loads in a catchment, can be achieved at least costs. Theoretically speaking, the least cost allocation of pollution abatement strategies is found if the marginal costs of the proposed measures are equal. The marginal costs of these abatement measures can for example be defined as the increase in total abatement costs when pollution loads are decreased by 1 ton or 1 kilogram per year. As long as marginal costs are not equal, it is theoretically possible to obtain the same level of pollution reduction at lower costs by shifting emission reduction from high cost measures to lower cost measures. In the WFD a CEA has to be carried out at catchment level.

Hence, also the spatial distribution of costs plays an important role and the question where measures should be taken. This is visualised in Box 4.1.

Box 4.1 Illustration of catchment-wide relationships

The Humber catchment in the North-East of England is the largest catchment in the United Kingdom and includes besides the Humber estuary the catchments of the river Ouse (C1) and Trent (C2). Sources of pollution (S) and potential measures (M) are also shown in the figure. Pollutants (e.g., nutrients or trace metals) are discharged and transported (T) into the Humber estuary directly and indirectly by point and diffuse sources. The rivers can be considered important sources of diffuse pollution. However, also within the estuary a number of important point and non-point sources are present (S7–S10). Different types of measures may be available to tackle these various sources of pollution.

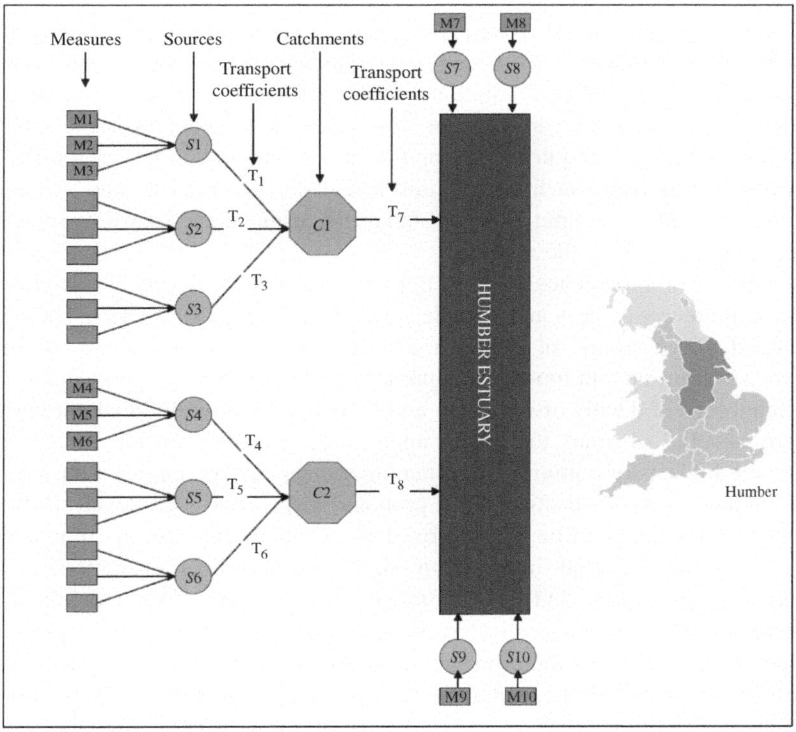

Source: Adapted from Cave *et al.* (2003)

The various steps distinguished in a CEA are described below:

Step 1: Define the environmental objective involved
Step 2: Determine the extent to which the environmental objective is met
Step 3: Identify sources of pollution, pressures and impacts now and in the future over the appropriate time horizon
Step 4: Identify measures to bridge the gap between the reference (baseline) and target situation
Step 5: Assess the effectiveness of these measures in reaching the environmental objective
Step 6: Assess the costs of these measures
Step 7: Rank measures in terms of increasing unit costs
Step 8: Assess the least cost way to reach the environmental objective

These steps are taken in sequence, but important feed-backs may exist between steps. As information becomes available about the problem, the source-effect pathway and possible solutions, the same step may be revisited several times. The outline of the various steps shows that carrying out a CEA is a multi-disciplinary exercise, requiring the input of and collaboration between different scientific disciplines, such as natural scientists, economists and technical engineers, but also the input of policy and decision-makers as they determine the scope and objective of the analysis.

A number of approaches are used in practice at varying levels of complexity, scale, comprehensiveness and completeness for carrying out a CEA. These are discussed, for example, in Zhang and Folmer (1995). A distinction is made between bottom-up and top-down approaches. The bottom-up approach focuses on technological details of measures and their impact on individual enterprises (micro level), whereas top-down approaches usually consider the wider economic impacts of pollution abatement measures and strategies, often without detailed technical specification of the proposed measures (macro level). Bottom-up approaches can also be characterised as technical engineering approaches, often including detailed information about the technical characteristics of production processes and only limited information about the financial engineering costs of emission abatement technologies. Top-down approaches on the other hand focus more on the economic relationships and consequences involved and less on the technical specification of measures. Examples of bottom-up approaches include ad-hoc approaches comparing a limited number of abatement technologies usually on a very local scale based on their engineering costs and emission reduction capacity (see Box 4.2) and the use of dynamic optimisation models where various abatement measures and technologies at enterprise or sector level are automatically prioritised with the help of linear programming (LP) techniques. Examples of top-down approaches are input-output and computable general equilibrium models. The inclusion of indirect effects depends on their relevance and the role they are expected to play in the decision-making procedure.

Box 4.2 Illustration of a cost-effectiveness analysis

In its most simple form, the cost-effectiveness analysis can be depicted as in the figure below. Various possible measures (M1–M8) are ranked in increasing order of their unit costs. Besides a different unit cost (reflected by a-h on the vertical axis), each measure also has a specific pollution (nutrient or metal) abatement or reduction capacity (reflected by A-H on the horizontal axis).

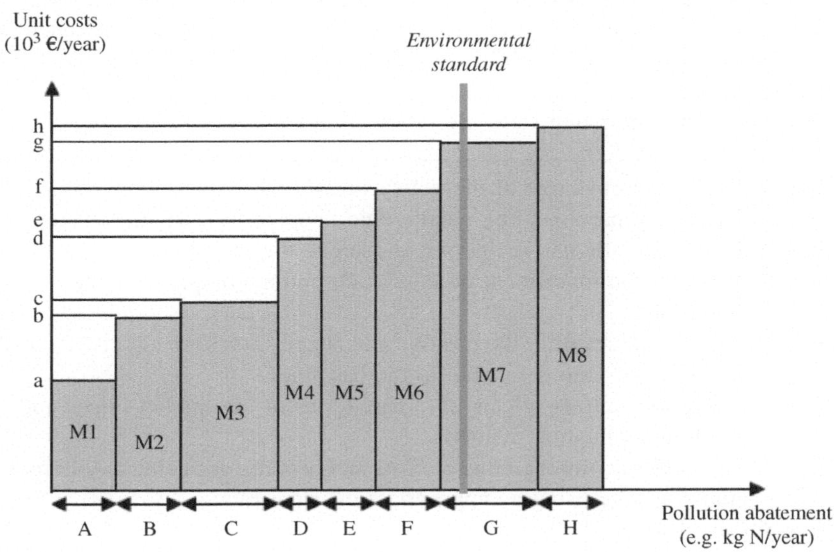

The environmental objective (standard) is represented by the vertical red line. The least cost option to reach the environmental standard is found by implementing measures M1 to M6 and, if possible, part of measure M7. The total costs of implementing these measures is found when taking the integral of the area M1-M8 from the origin until the environmental standard.

4.3. Cost-benefit analysis

Cost-benefit analysis (CBA) is carried out in order to evaluate and compare the economic efficiency of alternative actions. The benefits from an action are contrasted with the associated costs within a common analytical framework. To allow comparison of these costs and benefits related to a wide range of scarce productive resources, including water resources, measured in widely differing units, money is used as the common denominator. This is where most problems usually start for environmental policy and project appraisal since many environmental resources such as water are often not priced in monetary terms. For many goods and services provided by water resources, there is no market on which they are traded, and therefore no market price is available which reflects their economic value. There are, however, several economic valuation methods, which allow placing a value on non-marketed goods and services (see Section 4.4). The economic valuation of water uses and services compares the willingness to pay and opportunity costs of the goods and services supplied by water resources and the water environment. This means that a wide range of environmental goods and services can be explicitly recognised in the CBA process.

In general, the following steps are followed in a CBA:

Step 1: Define the objective of the policy measure
Step 2: Define the baseline, i.e., what would happen if no action is taken
Step 3: Define the alternative options to achieve the objective
Step 4: Quantify the investment costs of each option compared to the baseline
Step 5: Identify and quantify the positive and negative welfare effects of each alternative option compared to the baseline
Step 6: Value the welfare effects in money terms, using market prices and economic valuation methods
Step 7: Calculate the present value of costs and benefits occurring at different points in time using an appropriate discount rate
Step 8: Calculate the Net Present Value (NPV) or Benefit-Cost (B-C) ratio of each alternative option
Step 9: Perform sensitivity analysis
Step 10: Select the most efficient policy measure

Like CEA, carrying out a CBA is a multi-disciplinary process, involving expertise from different fields and the input from policy and decision-makers. While economists are involved in all steps, environmental expertise of many kinds is also needed, especially in steps 2 and 5. In order to ensure that the policy options are technically feasible, input from technical engineers is required especially in step 3, and often also in step 4 to specify the exact nature of the policy measure and estimate the associated investment costs. Policy and decision-maker input is essential when defining the objective, which the policy measures are supposed to achieve, and when defining the baseline and policy scenarios, including current policy. A key role of the economist in the whole process is to frame the issue and develop the CBA framework so that all relevant socio-economic stakes and the stakeholders are included and the multitude of environmental studies that need to be undertaken are working towards answering the following two questions:

(1) Is the action economically speaking worthwhile, that is, do the benefits outweigh the costs?
(2) And if so, which policy option yields the highest net benefit?

A CBA compares the costs and benefits of different policy options in monetary terms. The results of this analysis can be interpreted as a B-C ratio, that is total benefits divided by total costs, where a ratio larger than one indicates that the policy measure is economically beneficial, or as a NPV, that is the present value of the net benefits, where a positive NPV indicates a welfare improvement. Strictly speaking, only those costs and benefits are included in a CBA that can be quantified in monetary terms. However, it will hardly ever be possible to monetise all impacts all the time: those impacts that cannot be monetised are often left out of the analysis (see Box 4.3). Non-monetised impacts, if considered relevant, can nonetheless be included in a qualitative discussion accompanying the discussion of the CBA results.

Box 4.3 Illustration of a Cost-Benefit Analysis for Contaminated Sediment Clean-Up

The fictive cost-benefit balance sheet below reflects a classic situation in water policy and management where the costs of the water policy are estimated with a fair degree of accuracy, but the non-market benefits of the generated public goods and services are hard if not impossible to estimate in money terms and not included at all, as a 'pro memoriam' (PM) item or merely in qualitative terms.

Costs (million €)		Benefits (million €)	
Dredging costs	1,600	Market benefits increased commercial shipping	1,400
Storage costs contaminated sediments	1,200	Revenues from market sale of clean sand	250
Clean-up costs contaminated sediments	800	Ecological bebefits	PM
Total costs	3,600	Total benefits	1,650 +PM

While a textbook CBA requires that all impacts be monetised, in practice different approaches exist on how non-monetised impacts are included in the CBA. In some approaches they are listed as 'Pro Memoriam' items on the balance sheet, expressed in qualitative or quantitative form (Brouwer and Van Ek 2004). Pearce (1998) argues that in early CBA conducted in the UK, such impacts would have been either ignored entirely, left for a subsequent environmental impact analysis, or monetised only partly. This approach of monetising impacts where possible, and including them in another form where monetisation is not possible marks a deviation from the textbook ideal, but does not discredit the method as such.

An important distinction in CBA is that between a financial and economic CBA:

A *financial CBA*, also referred to as a cash-flow or a financial analysis, evaluates advantages and disadvantages of a project or measure in terms of the expenditures and earnings directly associated with its implementation for the investor. Originally devised for investment decisions, the tool can also be used to assess budgetary impacts of policies.

An *economic CBA* evaluates the costs and benefits of a project or measure in a broader sense, taking into account all positive and negative welfare effects, on the (national) economy as a whole and all relevant stakeholders. The costs and benefits addressed in an economic CBA may include indirect (second-order) effects and non-priced external effects on society and the environment. If such externalities are included in the analysis in monetary terms, it is usually also referred to as an *extended CBA*.

In practice, government policies are often evaluated primarily on the basis of their financial (budgetary) costs, as these can be assessed relatively easily. The calculation of economic costs and benefits, especially non-priced external environmental effects, is a more difficult task. This will be addressed in the next section. An economic or extended CBA is the more appropriate method for evaluating public water policies, since government interventions are often related to the provision of public goods, having an impact on society as a whole. Such impacts should consequently be valued and evaluated from a societal perspective, not the perspective of the investor only (e.g., government).

4.4. Non-market valuation methods

In economics, value is expressed as the degree to which people want to give up scarce resources, such as money or time, to acquire or retain something. Value exists in this sense only through the interaction between a subject (individual) and an object, and is therefore not considered an intrinsic quality of something (Pearce and Turner, 1990). As in other social sciences, the value people attach to something is based upon a hypothesised positive relationship between their observed behaviour or verbal responses and that value. In economics, an individual's value is usually revealed through market behaviour and measured in money terms by an individual's willingness to pay (WTP), for instance for a water quality improvement, or willingness to accept (WTA) compensation for a water quality decline. WTP and WTA cannot be used interchangeably. Although the WTP approach has become the most frequently applied and has been given peer review endorsement through a variety of studies (e.g., Arrow *et al.*, 1993), the question which concept is most appropriate in the context of water resources depends on the specific circumstances and the property rights regime associated with the specific water use.

Aggregated across those who benefit from natural resources and their services and who will hence be affected by any change in their provision level, including quality level, the aggregated WTP or WTA amount provides an indicator of their total economic value (TEV) (see Box 4.4). Environmental economists have introduced a taxonomy of this TEV, distinguishing between use values and non-use values, in order to account for the various reasons and motives people may have to value environmental change. Use values are associated with the actual or potential future use of a natural resource (e.g., drinking water, fish consumption, irrigation water). Non-use values are not related to any actual or potential future use, but refer to values attached to the environment and natural resource conservation based on considerations that, for example, the environment should be preserved for future generations or because plants and animals also have rights.

A range of valuation techniques exists for assessing the economic value of the functions performed by aquatic ecosystems (Box 4.5). Many water resources provide goods and services, which are not traded in markets and therefore remain un-priced. The valuation techniques developed over the past decades attempt to assess the relative economic worth of these goods and services using non-market valuation techniques. Depending on the nature of the specific environmental change in the water system, the presence of a market where the goods and services involved are exchanged and data availability, economic values can be estimated using direct or indirect market and non-market based valuation techniques. More detailed information about the underlying theory and practical implementation of these techniques can be found in general texts like Randall (1988), Braden and Kolstad (1991), Freeman (1993), Hanley and Spash (1993), Pearce *et al.* (1994) and Young (2005).

Box 4.4 Steps in the economic valuation of environmental goods and services

(1) Identification of the goods and services provided by water resources amenable to robust valuation
(2) Assessment of their provision (target) level, including quality attributes, compared to the baseline (reference) level of provision
(3) Identification of the groups of people in society (users and non-users) who benefit from the goods and services involved or who will be suffering a loss when they are removed, destroyed or degraded
(4) Identification of the possible values (use and non-use values) attributed to the goods and services involved by these groups in society
(5) Selection of the appropriate economic valuation technique(s)
(6) Estimation of the economic value of the change in provision level of the goods and services involved, accounting for substitution and income effects and other contextual factors

(7) Quantification of the 'market size', that is, the total population of beneficiaries over which the economic value is aggregated, accounting for possible distance-decay effects (people living further away may attach less value to the goods and services involved)
(8) Estimation of the total economic value

An important distinction is between valuation techniques, which estimate benefits directly and those, which estimate costs as a proxy for benefits. For instance, estimating damage costs avoided, defensive expenditures, replacement costs, substitute costs or restoration costs as part of an economic valuation exercise suggests that these costs are a reasonable approximation of the benefits that society attributes to the resources in question. The underlying assumption is that the benefits are at least as great as the costs involved in repairing, avoiding or compensating the damage involved. These techniques are widely applied due to their relative ease of estimation and availability of data, but it is important to be aware of the limitations in terms of the information they convey with respect to economic benefits. Investment by public bodies in conserving ecosystems may also represent a surrogate for aggregated individual WTP and hence social value. These 'public prices' paid for resources are also used to approximate the value society places upon them, for instance the costs of designating an ecosystem as a nature reserve.

Box 4.5 Valuation methods to estimate the socio-economic benefits of aquatic ecosystem functions

The choice of valuation method that can be employed to estimate the economic value of the good or service will often be up to the analyst to decide. While some methods are theoretically preferable to others, other 'second best' measures may often be easier to determine in practice. The choice of method is likely to depend, in part, on time, resources and data available for the investigation. An overview of appropriate valuation methods for different water system functions is given in the table below.

Aquatic ecosystem function	Direct use values	Indirect use values	Non-use values	Economic valuation method
Hydrological functions				
Flood water detention	√	√		HP, CV, CE, ADC, DE, RC
Ground water recharge/discharge	√	√	√	MA, PF, HP, TC, CV, CE, RC
Sediment retention	√	√		ADC
Biogeochemical functions				
Nutrient retention & export	√	√	√	MA, PF, HP, TC, CV, CE, DE, RC
Trace element storage & export	√	√	√	MA, PF, HP, TC, CV, CE, DE, RC
In situ carbon retention	√	√	√	ADC
Ecological functions				
Ecosystem maintenance	√	√	√	MA, HP, TC, CV, CE, RC
Food web support	√	√	√	MA, RC

Explanation:
MA: Market Analysis CV: Contingent Valuation RC: Replacement Costs (Shadow Project)
HP: Hedonic Pricing CE: Choice Experiment PF: Production Function
TC: Travel Cost DE : Defensive Expenditures ADC: Avoided Damage Costs
Source: Turner *et al.* (2004).

Market valuation means that existing market behaviour and market transactions are used as the basis of the valuation exercise. Economic values are derived from existing market prices for inputs (production values) or outputs (consumption values), through more or less complex econometric modeling of dose-response or damage functions. Examples include the economic value of fish, which is sold on a fish market (market analysis), the costs of replacing impaired environmental riparian functions such as nutrient retention and export through the installation of a wastewater treatment plant (replacement costs) or the costs of a water filter on tap water (avertive behaviour or defensive expenditures).

Where market prices exist for water resources, these may have to be adjusted for market distortions such as taxes or subsidies in order to obtain the real or shadow prices, but otherwise they are likely to provide a relatively simple means of assessing economic value. However, theoretically these estimations based on market prices are still not the same as the total economic value since they do not include the consumer surplus.

In the absence of market prices for water resources, the economic value of the goods and services provided by water systems can be estimated with the help of *direct* and *indirect* non-market valuation methods (e.g., Johansson, 1987; Mitchell and Carson, 1989; Freeman, 1993; 2003). Non-market valuation means deriving economic values in cases where such markets are non-existent or distorted. Direct methods (also called stated preference methods) refer to contingent valuation (CV), discrete choice experiments (CE), and contingent ranking (CR) techniques, where individuals are asked directly, in a social survey format, for their WTP for a pre-specified environmental change. WTP can also be measured indirectly by assuming that this value is reflected in the costs incurred to travel to specific sites (travel cost studies) or prices paid to live in specific neighborhoods (hedonic pricing studies) (also called revealed preference methods). The latter two approaches measure environmental use values through revealed preferences, while CV and CE are believed to be able to also measure non-use or passive use values through stated preferences. Of these methods, CV is probably the most widely applied method in contemporary valuation research (Carson *et al.*, 1995; Bateman and Willis, 1999). However, the validity and reliability of contingent valuation methods and their results are contested (e.g., Foster, 1997), especially in the context of non-use values. Moreover, conducting a contingent valuation survey can be a lengthy and resource-intensive exercise.

Irrespective of these criticisms, it is clear that valuation studies have a role to play in contemporary environmental policies, as they provide additional knowledge to support better decision-making (see www.aquamoney.org). It is important to apply and interpret economic valuation results in their appropriate context and to be aware of the pitfalls involved. However, this applies to most methods and techniques, in economics and any other field. Following best-practice recommendations is essential when applying the method in the field of water resources (for best practice recommendations, see for instance Arrow *et al.*, 1993 and Bateman *et al.*, 2002).

4.5. Benefits transfer

Instead of carrying out a new original valuation study, existing economic value estimates from previous studies can be used too. This is generally referred to as benefits transfer (Brouwer, 2000). Benefits transfer is a technique where the results of previous environmental valuation studies are applied to new policy or decision-making contexts. In the literature, benefits transfer is commonly defined as the transposition of monetary environmental values estimated at one site (study site) to another site (policy site). The study site refers to the site where the original study took place, while the policy site is a new site where information is needed about the monetary value of similar benefits.

In the field of environmental valuation, benefits transfer has been applied extensively in various contexts, ranging from water quality management (e.g., Luken *et al.*, 1992) and associated health risks (e.g., Kask and Shogren, 1994) to waste (e.g., Brisson and Pearce, 1995) and forest management (e.g., Bateman *et al.*, 1995). Costanza *et al.* (1997) extrapolated the monetary values of existing valuation studies to the flow of global ecosystem services and natural capital, and thereby raised a number of questions as well as heavy criticism about the reliability of benefits transfer (Box 4.6).

The most important reason for using previous research results in new policy contexts is that it saves a lot of time and money. Applying previous research findings to similar decision situations is a very attractive alternative to expensive and time consuming original research to inform decision-making. In practice, several approaches to benefits transfer can be distinguished, which differ in their degree of complexity, data requirements and the reliability of the results.

The first two approaches are most frequently applied, as they require relatively little data or expertise, and are usually not very time consuming. First, the average WTP value from another study can be used to predict the economic value of the benefits involved at the policy site. The old study ideally focuses on the same environmental goods and services, but may have been carried out at a different location or at a different point in time. A second approach is to estimate an average WTP value from the mean value estimates of several similar studies.

Box 4.6 Transfer errors found in the benefits transfer literature

Although benefits transfer is used extensively in practice, relatively little published evidence exists about its validity and reliability. Table 4.1 gives an overview of water related studies, which tested the reliability of the transfer of WTP values. Although not complete, the table shows that most studies test the reliability of transferring contingent values. Three studies investigate the transferability of travel cost studies. The estimated benefits in these studies are related to different types of water use, such as recreational fishing, boating or other recreational water use. Also the study by Bergland *et al.* (1995) and Parsons and Kealy (1994) look at water quality improvements for recreational use. The last column presents the range of transfer errors found in these studies. So, a transfer error of 50% means that transferring the value from a study site to a policy site can be 50% higher or lower than the 'true' value at the policy site. The range of transfer errors refers to the transfer of average WTP values and WTP functions.

Table 4.1 Errors found in water related economic valuation studies testing benefits transfer

Study	Valuation method	Estimated benefits	Transfer errors (%)
Loomis (1992)	Travel cost	sport fishing benefits	5–40
Parsons and Kealy (1994)	Travel cost	water quality improvements	1–75
Loomis *et al.* (1995)	Travel cost	water based recreation	1–475
Bergland *et al.* (1995)	Contingent valuation	water quality improvements	18–45
Downing and Ozuna (1996)	Contingent valuation	saltwater fishing benefits	1–34
Kirchhoff *et al.* (1997)	Contingent valuation	white water rafting benefits	6–228
Brouwer and Bateman (2005)	Contingent valuation	flood control benefits	4–51

From Table 4.1 it is difficult to say how large the errors can be expected to be when using existing economic value estimates in new decision-making contexts. In some cases they can be very low, in other cases they can be as high as almost five times the value, which would have been found if original valuation research was carried out. No distinct differences can be found when comparing transfer errors for contingent valuation and travel cost studies. The errors reported in Table 4.1 have to be considered in the light of the purpose for which the user (policy or decision maker) wishes to use previous valuation results. In some cases a transfer error of 50 percent may be considered too high, in other cases such an error may be acceptable. User acceptability of the error will depend on subjective judgement by the user self, the purpose and nature of the cost-benefit evaluation and the phase of the policy cycle in which the evaluation is carried out.

Source: adapted from Brouwer (2000).

A third approach is to use an average WTP value, which is adjusted for one or more factors that are expected to influence the new value estimate at the policy site. For instance, average WTP may be adjusted for differences in income levels between the study and policy site. This means that in the old study information has to be available about the way income influences WTP.

Fourth, the entire WTP function from an original study can be used to predict mean WTP at the policy site. The estimated coefficients in the WTP function are multiplied by the average values of the explanatory factors in the new policy context to predict an adjusted average WTP value. This approach is expected to be more robust than the transfer of unadjusted average WTP values, since effectively more information is transferred (Pearce *et al.*, 1994). However, it is also more data intensive as secondary information about all relevant factors has to be collected.

A fifth approach is to use a WTP function, which has been estimated based on the results of various similar valuation studies. The WTP function is in this case estimated on the basis of the summary statistics of different studies. This approach is usually referred to as meta-analysis.

A sixth approach is the use of a WTP function – either based on a single or multiple studies – where the coefficient estimates are adjusted when transferring the estimated value function based on prior knowledge about the impact of the explanatory factors on WTP in the new policy situation.

A number of criteria have been identified in the literature for benefits transfer to result in reliable estimates (e.g., Desvousges *et al.*, 1992; Loomis *et al.*, 1995; Brouwer, 2000):

(1) good quality data
(2) similar environmental goods and services
(3) similar sites where these goods and services are found
(4) similar populations of beneficiaries
(5) similar market constructs
(6) similar market size (number of beneficiaries)
(7) similar number and quality of substitute sites where the environmental goods and services are found.

Study quality is an important criterion, which can be assessed in a number of ways. Above all, one can look at the internal validity of the study results, that is, the extent to which findings correspond to what is theoretically expected. This internal validity has been extensively researched over the past three decades in valuation studies. Studies should contain sufficient information to assess the validity and reliability of their results. This refers, among others, to the adequate reporting of the estimated WTP function, including the applied statistical techniques and the definition of variables.

Thus, while benefit transfer provides a quick and cheap alternative to original valuation research, some conditions must be met if it should provide reliable results. Above all, the local circumstances and conditions in the new decision-making context need to be close enough to the ones prevailing in the original research. The risk of obtaining misleading results may be controlled and reduced by integrating more explaining variables into the transfer, however this also increases the data requirements and the complexity of the analysis. Also, the possibilities of conducting a sound and reliable benefits transfer hinge on the number, quality and diversity of valuation studies available – the larger, the better and the more diverse the existing set of studies is, the more likely there will be a primary study that is close enough to the policy site for results to be transferable.

4.6. Multi-criteria analysis

A Multi-Criteria Analyis (MCA) is a structured approach used to determine policymaker, stakeholder or public preferences for alternative policy measures, where policy measures may pursue multiple objectives. It is used to structure a policy problem in terms of possible alternative solutions evaluated under different multiple criteria.

MCA is applicable to cases where a single objective or single criterion approach like CBA, which compares options based on their economic efficiency only, is insufficient. Instead, a MCA accommodates a range of social, environmental, technical, economic, and financial criteria. MCA is especially suitable in situations where multiple environmental and social objectives are pursued, which cannot easily be expressed in monetary terms. MCA techniques are often integrated with participatory approaches, involving different policy and stakeholder groups (see for example the UNEP 'Value Base Assessment' procedure and use of MCA in relation to transboundary water management).

One setback of MCA in these participatory approaches is that the method is often difficult to use for lay people. Most of them require an expert to explain how the method works, help users to define options, criteria and weights, as well as to choose the appropriate aggregation procedure. An important question is which approach the experts and the end-users of the results feel most comfortable with when making effects comparable and commensurable, that is through economic valuation of effects or standardisation and weighting procedures of impacts based on the statistical standardisation of impacts and criteria weights as in MCA.

The general steps in MCA are described below:

Step 1: Define the problem and the main objectives of the policy measures
Step 2: Define the alternative options to achieve the objectives
Step 3: Identify the criteria to evaluate each option compared to the baseline, including their measurement in monetary and non-monetary units
Step 4: Quantify the effect or performance of each alternative policy measure on each criterion compared to the baseline
Step 5: Complete the effects table or performance matrix
Step 6: Standardize the effect scores in the performance matrix
Step 7: Attach weights to the criteria
Step 8: Aggregate the weighted effect scores to a global score for each policy measure
Step 9: Perform sensitivity analysis
Step 10: Select the most preferred policy measure based on the identified criteria

As for a CBA, MCA starts off by identifying and framing the problem to be assessed, and formulating the objectives that are to be reached through the policy intervention. Such objectives may include environmental targets as well as social cohesion, employment generation or income distribution. This is followed in the second step by the identification of different options to achieve the objectives, a step that may involve the participation of different stakeholders and the wider public. If too many potential measures are identified, an initial screening may be helpful to pre-select the available options. This screening can take place based on a qualitative MCA.

A third step is to identify the criteria and possible sub-criteria against which the different options should be evaluated, along with their suggested indicators and measurement units. The identification of the criteria will be closely related to the objectives identified in the first step. The criteria can be measured in monetary and non-monetary terms, but they may also be qualitative in nature. An important difference between MCA and CBA is that the former is able to relate information in different forms to multiple objectives measured through multiple criteria. MCA incorporates environmental, economic and social effects as in a CBA, but trades them off against each other in a different way, namely based upon multiple criteria measured in different units. While the weighing procedure and trade-off of different criteria is more complex for a MCA than for a CBA, it means that a MCA does not require the use of monetary valuation studies in the same way that a CBA does.

In the following step, the policy options identified in step 2 are assessed in terms of the criteria defined in step 3. To do this, the policy options are evaluated compared to a common pre-defined baseline scenario, mostly a 'business as usual' scenario. The performance of the different options against the baseline scenario are then presented in a scorecard, also referred to as an effects table, performance or impact scoring matrix (Box 4.7). A score is provided for each option against each criterion, measured in different units. These scores are then standardized between 0 and 1. For this, several techniques are available depending on the starting point and end point of the effect scores and the expected distribution of intermediate values.

Box 4.7 Illustration of a performance or impact scoring matrix for different flood management schemes

MCA Performance Matrix
Flood Management Schemes

Criteria	Measurement unit	Do nothing	Dike defences	Reconnect floodplain	Partial retreat	Full retreat
Safety (return period)	probability	1/50	1/1250	1/1000	1/500	1/250
Investment costs	10^6€	0	428	8353	5350	6487
Operation costs	10^6€	0	0	250	305	293
Damage costs	10^6€	9472	0	0	0	0
Homes affected	Number	0	0	2290	320	1540
Agricultural land	ha	0	0	15255	2840	10130
Industrial estates	ha	0	0	310	90	230
Sand extraction	10^6m^3	0	0	74	26	25
Nature development	%	0	0	60	55	70
Spatial quality	+/-	0	0	–	+	++

In step 7 the different criteria are prioritized by attaching weights to them. These weights can be assigned by the analyst, the decision maker, or they can be elicited from stakeholders through some form of participatory consultation process. Another option is to involve a panel of experts (Delphi method). A direct analysis of the performance matrix through policy maker screening of alternative options is of course also possible. The determination of weights is crucial, at the same time it also introduces an element of subjectivity into the decision making process. The weighting procedure basically corresponds with the valuation step in a CBA. A sensitivity analysis is essential to assess the effect of different weights on the outcome of the analysis.

For the calculation of a global effect score, the weighted scores have to be aggregated. For this aggregation a number of approaches have been developed. The following two main approaches can be distinguished (DETR, 1998): the compensation method and the outranking method. The compensation method is the best-known method where a global score is calculated for each policy measure in the form of a weighted arithmetic average of the scores attributed to that measure for the different criteria. The method called 'compensatory' because the calculation of the weighted average makes it possible that low scores on one criterion are compensated by high scores on another. The outranking method on the other hand is used if not all criteria are considered commensurable, and therefore no weighted aggregate score can be calculated. The analysis is based on paired comparisons of measures with respect to individual criteria. All possible comparisons are analyzed and synthesized in terms of 'policy measure A is at least as good as policy measure B in relation to a majority of criteria, without being altogether too bad in relation to the other criteria' (DETR, 1998).

The final step in a MCA is to prioritise, compare and order the alternative policy measures based on the evaluation results. Different methods can be used to this end. Depending on the method chosen, a MCA can be used to identify a single most preferred option, to rank options, to short-list a limited number of options for subsequent detailed appraisal, or to distinguish acceptable from unacceptable options. The prioritisation of different options should also involve a sensitivity analysis, investigating how the ranking order of the options changes if key parameter values and criteria weights were to be changed.

4.7. Integrated river basin models

The complexity of interactions between water and economy is increasingly translated into formal (mathematical) reduced-form models by linking relevant hydrological and biogeochemical structures and processes to economic 'laws' of supply and demand of water services. Historically, these models have been developed by water and civil engineering scientists, focusing on single and multiple objective decision-making and trade-offs (e.g., Dudley, 1972; Braat and

Lierop, 1987; McKinney and Cai, 1996; Andreu *et al.*, 1996; Cai and McKinney, 1997; Rosegrant *et al.*, 2000). Water use in agriculture (irrigation) has been the prime focus of many of these models given the fact that agriculture is the largest water consumer in the world (FAO, 2004). These models often include a detailed coupled hydrological and geological module – and in some cases also hydraulic and biogeochemical modules – to control for the hydro-geological heterogeneity in a basin area and are based on node networks.

Node networks are graphical delineations of water flows and stocks in a watershed or basin into different water quantity and/or quality balance (monitoring) stations linked to specific water demand and supply. For each geo-referenced node, a water demand and supply function is estimated based on the geographical unit's hydro-geological and biogeochemical characteristics. In the case of agriculture, the demand and supply functions are based on an agronomic model, such as a crop yield function, which depends on factors like soil, crop acreage, rainfall, crop evapotranspiration and irrigation system characteristics. Economic behavior is usually included through a profit maximization objective function, where fixed and variable production costs are subtracted from the yield benefits subject to the natural resource constraints such as land and water availability. The latter is obviously dependent on the hydro-geological conditions involved, including water supply and water quality constraints.

The key to integrated water modelling is that water systems perform economic functions and can be used as a source and a sink. After some degree of transformation, water is used as a source for economic consumption like drinking or recreation, and in economic production as an input factor in crop and food production, energy, paper, textile or metal production. At the same time, water is also used as a sink for the by-products of economic production and consumption processes such as the emission of polluting substances into surface and groundwater bodies. The interaction between the water and economic realm works both ways: water is transformed for economic use and the impact of economic use on water availability and quality consequently has implications in both the short and long term for the transformation process to adapt water for economic use. It is this latter dynamic feedback mechanism, which is often ignored and omitted in integrated water models.

Following Braat and Lierop (1987), a distinction is made in the literature between two different approaches to integrated hydro-economic model development, i.e., (1) models which allow for an effective transfer of information from one component to the other: the compartment or modular approach and (2) the holistic approach.

In the modular approach a connection is built between the hydrological and economic model, and output data from one module usually provides the necessary input for the other. In principle, the modules operate independently of each other and systems of equations are solved in an exogenous way (input variables from one model into the other are exogenous). In holistic models, variables that are exogenous in a modular approach are solved endogenously in a system of equations (Cai and Wang, 2006). However, under the modular approach, a loose connection exists between the different hydrologic and economic components. The various sub-models can be very complex and the main problem is to find the right transformation of data and information between sub-models. In the holistic approach there is one single unit with both the hydrologic and economic component tightly interwoven in a consistent endogenous model (see the example in Box 4.8). In order to be able to solve the complexity of simultaneous equations the different components have to be represented in a very simple way (McKinney *et al.*, 1999).

So, whereas information transfer between the various compartments or sub-models is one of the most important technical obstacles in the modular approach, the most important issue in the holistic approach is to find one single technique and denominator for the variable quantities and represent both the simplified hydrological and economic component in a meaningful way.

Important limitations to integrated modelling include (McKinney *et al.*, 1999):

- Hydrological models are often based on simulation techniques, whereas economic models usually use optimization techniques.
- Different spatial scales: water bodies, water sheds or basins usually are the geographical unit in hydrological models, while economic models often refer to administrative boundaries of a region (county, province, state) or a country as a whole.
- Different time intervals and horizons: time scales in hydrological models often refer to days, months or seasons (summer and winter), while in economic models the time scales (intervals and horizon) are usually larger than that (years).

Box 4.8 Illustration of a holistic hydro-economic model: AQUATOOL

The Hydraulic Engineering and Environmental Department of the Polytechnic University of Valencia, by means of a collaboration agreement with the Júcar River Basin District has developed AQUATOOL for assessing the marginal opportunity cost (MOC) of water resource use at the basin river scale. The MOC is defined as the cost of not being able to allocate one unit of water (m^3) to its economically most profitable use. This can be interpreted as a resource 'scarcity cost'. AQUATOOL is able to account for spatial resource availability, storage capacity, water losses due to evapotranspiration, return flows, surface and groundwater interactions, and willingness-to-pay of the various demand units. To obtain the resource cost at a certain location and time (for instance during a given month), a specific water volume is added or subtracted at the location of interest. The model then calculates through a new resource allocation pattern, using the modelled allocation rules, yielding a total economic benefit in the new situation. The benefit/volume ratio under the new conditions of resource allocation is an approximation of the marginal resource cost, reflecting the economic cost of water scarcity.

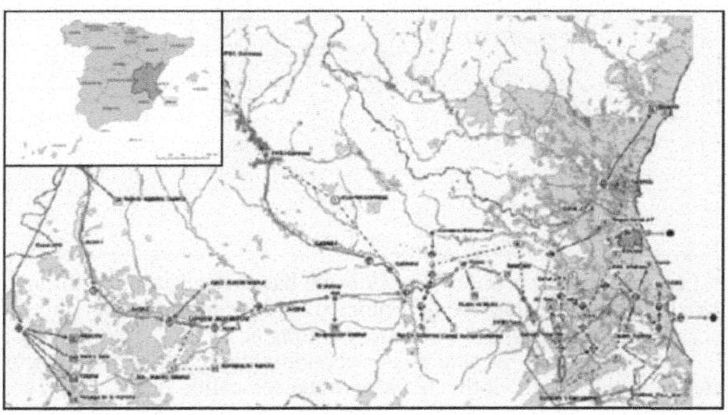

Source: Jucar River Basin Provisional Article 5 Report Persuant to the Water Framework Directive, Ministerio de Medio Ambiente, Confederación Hidrográfica del Júcar.

Besides the distinction between holistic and modular approaches, integrated hydro-economic models can furthermore be classified as (1) water use models, (2) water system models and (3) effects models. An overview of these different model types is given in Figure 4.2.

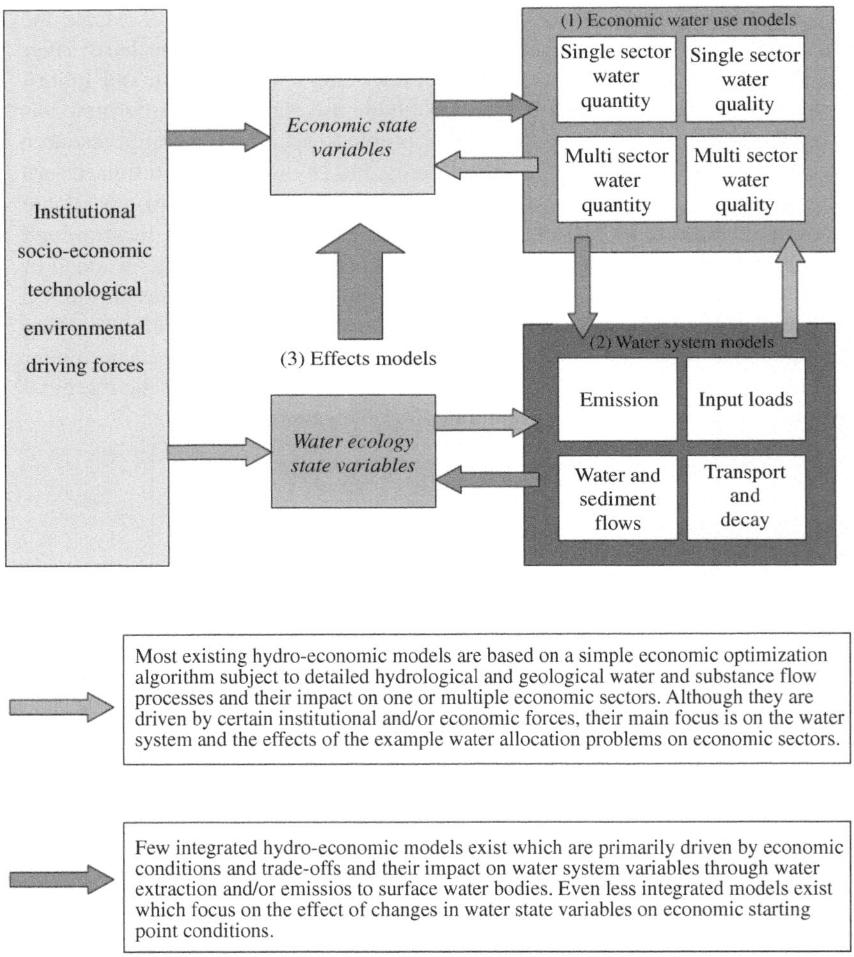

Figure 4.2 Different types of integrated hydro-economic models

Most integrated hydro-economic models are water system models, i.e., models based on detailed node networks of water and substance balances throughout the river basin, linked to an economic activity through a demand function. This demand function often depends on fixed (exogenous) technical input-output parameters of the economic production process involved (e.g., irrigation demand from agriculture), and reflects at best a partial economic equilibrium system of demand and supply equations.

Water use models are integrated models based on economic demand (consumption) and supply (production) functions, which are related to different forms of water use where water is an essential input in consumption and production processes. As most existing water system based hydro-economic models, water use models often reflect partial equilibrium states. Examples include input-output models of direct and indirect water use (e.g., Velázquez, 2005). More recently computable general equilibrium models are also used to link economic activities to water use, including water extraction and water pollution through the emission of polluting substances to water bodies (Brouwer and Hofkes, forthcoming).

Effects models incorporate changes in the water system and their effect on the economic system. Most existing models are partial equilibrium models, based on production function approaches, where water is one of the input factors and changes in the availability or quality of water needed for the production process at hand is assessed through estimated dose-effect relationships, which are subsequently related to market prices in order to derive a marginal value for water use.

Sometimes also 'meta-models' are distinguished as a separate class of modelling tools. Generally, meta-models are used to construct and develop general frames around specific problems analyzed with the help of a variety of data, expeert judgements and models. An example is given in Box 4.9. Meta-models integrate simulation results from sub-models (e.g., an economic optimisation model and a water quality simulation model) in a cause-effect framework. They are called meta-models because they only include the results of underlying models, where underlying model response surfaces are summarised for example in conditional probability distributions.

Finally, in the context of the WFD, a distinction can be made across European Member States between modelling efforts focusing on water scarcity, water allocation problems and the estimation of shadow prices and opportunity costs of alternative water use (especially in southern Europe), and modelling efforts focusing on water pollution, applying primarily bottom-up linear programming approaches to assess the cost-effectiveness of water quality measures (especially in northern Europe). An overview of existing integrated hydro-economic models is provided in the annex to this paper, based on the international workshop about the potential role of integrated river basin modelling in the WFD held in Amsterdam in November 2005 (www.ivm.falw.vu.nl/watereconomics).

Box 4.9 Illustration of the use of 'meta-models'

The Norwegian Institute for Water Research (NIVA) has evaluated the use of Bayesian belief networks for cost-effectiveness and benefit-cost analysis under the Water Framework Directive. A Bayesian network approach was used to conduct decision analysis of nutrient abatement measures in the Morsa catchment in South Eastern Norway. Bayesian networks and influence diagrams (see the Figure below) were used as a 'meta-modelling tool' for structuring and combining the probabilistic information available in existing cost-effectiveness studies, eutrophication models and data, non-market valuation studies and expert opinion. Hugin Expert software (www.hugin.com) was used to evaluate the cost-effectiveness of mitigation measures and disproportionality of their costs relative to the benefits. The most common nutrient mitigation measures in the catchment like precautionary tillage practices, vegetation buffer strips, sedimentation dams and individual waste water treatment were shown to have little effect on the lake water quality when uncertainty in the models, data and expert opinion underlying the driver-pressure-state-impact chain was modelled explicitly. Bayesian belief networks are well suited to integrate natural and social science model results within a DPSIR framework, and for evaluating the effects of joint model uncertainty on alternative programmes of measures. Despite the Morsa catchment being one of the most intensively studied catchments in Norway, the application was limited among others by the lack of adequate uncertainty analysis in existing studies and models, and the unfamiliarity of planners with the probabilistic recommendations made by researchers.

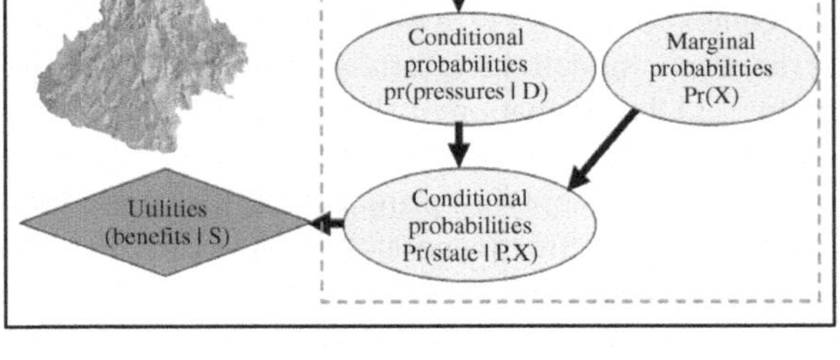

Source: Barton *et al.* (2006). Using belief networks in pollution abatement planning. Example from the Morsa catchment, South Eastern Norway. NIVA Report No. 5213-2006, Norwegian Institute for Water Research.

One of the workshop outcomes was that integrated models have a role to play in the economic analysis in the WFD in article 5 (river basin characterization), article 11 (selection of a cost-effective programme of measures), article 9 (economic instruments), and article 4 (disproportionate costs).

In order to be able to predict the impact of water policy changes to reach the environmental WFD objectives on both the water and economic system, the causal relationship between water and economic production and consumption processes has to be established first, for example through the article 5 reports and the use of integrated river basin information systems, as described in Section 3.1. This includes for the selection of a cost-effective programme of measures the physical dose-effect relationships in order to be able to assess the impact of end-of-pipe or process-integrated measures on the water system. Furthermore, if large-scale interventions are foreseen, the economic impacts may not be limited to the sector involved, but cause a chain reaction across interlinked sectors. For example, if measures in agriculture result in substantial changes in inputs and outputs, the whole agribusiness may be affected, not only the primary production of agricultural produce. For the assessment of the concept of disproportionate costs, the estimation of both the direct and indirect costs will be relevant. Hence, there is a clear need for consistent (dis)aggregation procedures to assess the sector-wide and in some cases also the regional and national economy wide impacts of the WFD implementation (see Figure 4.3).

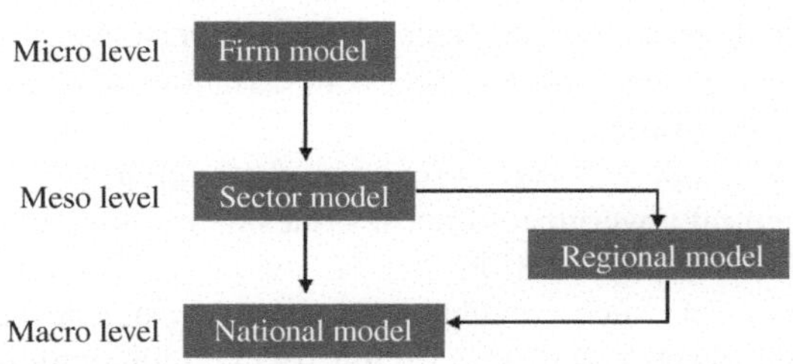

Figure 4.3 Consistency in aggregation and scaling procedures of economic models

The same applies to the introduction of economic instruments for cost recovery purposes. The effectiveness of economic instruments in changing economic behaviour (e.g., demand for a water service) depends on price and income elasticities of demand, which have to be estimated as information about such elasticities is usually not ready available. If such an introduction furthermore results in changes in demand and supply, and hence relative market prices, a modelling approach is needed to assess the effect of these changes on the relevant economic state variables instead of the usual 'ad hoc' approach to assess the financial engineering costs of technical measures.

4.8. Economic instruments

Economic instruments are used to correct market failures. Market failures usually imply that the market structure and corresponding market performance impede the efficient allocation of resources, goods and services across economic agents or sectors. Market prices may not reflect the real value of the resources involved due to for example the existence of monopolies and oligopolies (e.g., cartels) or the public good nature of the goods and services involved. Water resources often exhibit such public good characteristics (see Section 3.5). An economic instrument helps to ensure that economic agents or sectors account for the real value of water resources in their activities. For instance, the introduction of a water pollution levy may ensure that companies discharging wastewater into surface waters pay for the environmental cost they impose on society based on the Polluter Pays Principle. The external cost to society is in this way internalized through the economic (market based) instrument.

On the other hand, economic instruments may also be used in well-functioning markets to adjust for fairness and equity considerations to ensure a fair and equal distribution of resources across economic agents or sectors. Subsidies are a classic example of such 'market failure'. Economic agents or sectors may be compensated for the fact that they represent a relatively weak or vulnerable group in society with less access to certain resources than others and is hence less well-endowed.

The traditional approach to correct market failures is environmental regulation. The central government imposes rules on resource use and allocation and hence commands and controls the behaviour of economic agents and sectors. Examples of such regulatory measures include the setting of environmental standards, the prescription of certain pollution abatement technology in production processes or product specification (e.g., content of washing powder or paint solvents). Non-compliance usually results in legal action and financial penalties.

Economic instruments have a number of advantages compared to the traditional 'command and control' approach (www.economicinstruments.com):

(1) *Cost minimization*: pollution abatement will occur where it is cheapest to do so. The decision where and how to abate is left to the economic actors and sectors, who know best how to reduce pollution at the lowest costs possible. Flexibility in meeting environmental objectives is therefore a key

characteristic. Firms with high pollution levels are penalized most and hence expected to be encouraged to look for cheaper environmentally friendly production methods, while firms with relatively low pollution levels are in a sense rewarded for their environmentally best behaviour.

(2) *Reduction of bureaucracy*: resource use and allocation is left to the market. Market prices signal economic efficiency and result in optimal use of the limited available resources. The self-regulating power of markets is expected to reduce administrative implementation and control costs. On the other hand, some level of monitoring and control may always be necessary in situations where there is asymmetric information and the economic agent may have an incentive to 'cheat'. Reduced bureaucracy (rules and administration) may also enhance environmental performance by reducing economic agent aversion against existing bureaucracy (imposed rules and regulation) and hence making the implementation of best available pollution control techniques easier and more attractive for economic agents and sectors.

(3) *Improved longer-term environmental performance and dynamic efficiency*: economic instruments are believed to stimulate longer-term environmental performance by encouraging economic agents and sectors to continuously search for the lowest pollution abatement cost possible. Contrary to environmental standards set under 'command and control' regulation, there is always a financial incentive in place to pollute less.

Other benefits include the optimal use of specialized knowledge and information about how to best achieve the environmental objectives and adaptive capacity to changing environmental and economic conditions.

The most important economic instruments are charges and taxes, emissions trading systems and subsidies. Auctions and tenders may also be useful to minimize public expenditures for water resource management and increase competitiveness in the allocation of public funds. These market-based instruments are used especially for conservation contracts (e.g., Stoneham *et al.*, 2003). Examples include wildlife, soil erosion and riparian management contracts where contracts are concluded through closed or open bidding rounds and procedures.

There do not exist any 'hard' guidelines for the selection, design and implementation of economic instruments. Their selection and design depends on the specific circumstances and conditions. It is advisable to involve policy-makers, stakeholders (economic agents and sectors involved) and economic advisors (www.economicinstruments.com). Improvement of environmental performance and cost minimization are the most obvious selection and design criteria, and correspond with the WFD requirement of selecting cost-effective measures. For this, it is important to have sufficient insight in the technology base and cost structure of the participating economic agents and sectors (often considered a key advantage of tenders is the disclosure of this type of information). Other generic criteria include (Bohm and Russell, 1985; Bovenberg *et al.*, 1991; OECD, 1997):

- Effectiveness in reaching the environmental objectives
- Ease of monitoring and enforcement
- Administration and compliance costs
- Dynamic innovation effects
- Wider economic effects on competitiveness and employment
- Flexibility in the face of economic change
- Political acceptability
- Soft effects like awareness, capacity building, diffusion of information.

A practical problem is that many if not most of the criteria mentioned are difficult if not impossible to measure *ex ante* due to the fact that experiences with market based instruments in water resources management is limited. In situations where experiences with market based instruments are limited or absent, the adoption of an 'adaptive management' approach seems to be a sensible way, experimenting first on a small scale (including all relevant parties involved), and modifying and extending the use of market based instruments as the water manager's knowledge and experience base increases based on careful monitoring of the impacts of the instrument.

One way of evaluating the effectiveness of an economic instrument is through the concept of price elasticity of demand. The price elasticity of demand measures how water use as an intermediate or final product changes as a result of a unit price change. The empirical evidence shows that water use is inelastic, meaning that a unit price change has little effect on water use, for example by domestic households and agriculture. An important question is how expensive water has to become before water use becomes elastic (in other words: what is the exact shape of the demand curve for water).

A specific problem here furthermore arises when trying to assess the costs of the economic instrument. Three main cost types can be distinguished:

(a) The adaptation costs of a change in economic behaviour (usually incurred by the economic agents and sectors). This includes investment costs by economic agents and sectors in pollution abatement technology (end-of-pipe or process integrated technology), operation and maintenance costs, and other adaptation costs. These costs are net of any cost savings due to more efficient production processes (e.g., lower energy costs).

(b) The costs associated with charges, levies and taxes paid and subsidies received (usually incurred by the economic agents and sectors, but also the government). Also here it is important to check for possible net effects if economic agents and sectors are compensated for possible income losses through for instance income subsidization.

(c) The transaction costs associated with the introduction and implementation of the economic instrument such as monitoring and administration costs of permit emission, registration, inspection, sanctions etc. (incurred by the economic agents and sectors and the government). This latter category is often the most difficult category to measure in practice. The costs associated with the introduction and implementation of market-based instruments are often hidden in general overhead and administration costs.

Finally, in the case of water resources management, the identification of the specific situation characteristics is crucial in the design of the economic instrument. This includes:

(i) The nature and complexity of the water resource management problem such as the degree of pollution or water abstraction, the spatial scale of the problem (e.g., transboundary), the temporal scale of the problem (time gap between the moment of pollution and its manifestation in surface or groundwater quality).

(ii) The nature and complexity of the driving forces and pressure sources, including the availability of alternative production technologies and the measurability of the polluting processes.

(iii) The nature and structure of the target group(s), including point and diffuse sources of pollution, their size, spatial distribution and degree of organization.

(iv) The nature and structure of the market, including market imperfections, which reduce the price elasticity of water use (e.g., subsidies), competitiveness and mobility of input resources (e.g., water transfers).

Obviously, the success of the implementation of economic instruments largely depends on their design and the cost-effectiveness of the selected instrument in reaching the environmental WFD objectives. This is in turn determined by the intrinsic nature of the instrument and its ability to address and account for the generic and situation specific characteristics discussed in this Section.

5. SUMMARY AND CONCLUSIONS

This Guidance Report aimed to provide an overview of the key issues in the economic policy analysis underlying the WFD. The report's target group is economic practitioners, water policy advisors and consultants with some familiarity of economics working on WFD implementation. Although an attempt was made to write the report in understandable, non-technical terms, the nature of the issues raised in this report inevitably result in the use of jargon. Some knowledge and understanding of economics is therefore expected to be necessary when reading the report.

The following five key issues were identified, described and elaborated: integrated river basin information systems, the selection of a cost-effective program of measures, the assessment of disproportionate costs, the estimation of environmental and resource costs, and the design of incentive compatible water pricing mechanisms. Where possible international examples were used to illustrate the issues, but practical examples directly related to WFD implementation (besides the available article 5 descriptions) are still scarce. Following the discussion of these key issues, economic textbook overviews were given of available methods, tools and instruments to address the economic policy analysis challenges in the context of the WFD. These include cost-effectiveness analysis, cost-benefit analysis, multi-criteria analysis, integrated hydro-economic modelling and economic instruments.

The key issues are closely linked and interrelated. Available information about causal interactions, including dose-response relationships of current water use, where economic activities exert pressures on water availability and quality, provide the basis for the economic assessment of programs of measures aimed at changing current water use to achieve good water status by 2015. Such programs will include modified and new economic instruments for water pricing, adequately accounting for the environmental and resource costs of water use patterns based on combinations of economic valuation methods and shadow price modelling. Whether the estimated costs and the distribution of these costs across different public and private water users is considered disproportionate depends on the prices these water users are asked to pay, their willingness to pay and their ability to pay.

Surrounding each of the key issues identified in this paper is the issue of *scale* and *uncertainty*. Standard economic methods and models are not directly applicable to the issues raised here due to the specific characteristics of the water economy. The level of analysis at river basin scale introduces a whole range of new challenges, like up/downscaling procedures of economic costs and benefits due to upstream-downstream interactions and the correct interpretation of the economic results. Carrying out the economic analysis at (international) river basin scale instead of local water body or sub-regional level is expected to be economically beneficial, because it may result in substantial cost savings. Preventive measures at source upstream can be much cheaper than damage repair and remediation downstream and corresponds with the policy principle of prevention rather than cure. Moreover, the estimated costs of selected programs of measures may be considered disproportionate at local level (e.g., in the order of magnitude of hundreds of thousands of euros or maybe even a few million euros), but negligible at national level (e.g., less than 0.001% of GDP) or international river basin level if there are effective economic-institutional structures in place for fair cost sharing.

In the case of disproportionate costs, the (un)certainty of reaching the environmental WFD objectives by 2015 or maybe later, the (un)certainty of the associated direct and indirect economic costs and benefits, and the (un)certainty surrounding the allocation of the costs across different public and private water users through a variety of financial instruments and payment structures plays an important role in the WFD decision-making process too. For the assessment and evaluation of different types of uncertainties surrounding the economic components in the WFD, the interested reader is referred to Brouwer (2005). Addressing these issues in more detail here is beyond the scope of this paper, but they clearly require and deserve more attention in the economic analysis underlying the WFD and future pan-European research programs than currently is the case. The role of integrated hydro-economic modelling in particular is expected to play an increasingly important role as a more formal organizing framework of available knowledge and information, including qualitative expert judgment, for the future design of cost-effective programs of measures and water pricing instruments accounting for possible indirect economic impacts and environmental and resource costs and benefits.

6. REFERENCES

Adamus, P.R. and Stockwell, L.T. (1983). *A Method for Wetland Functional Assessment: Volume 1. Critical Review and Evaluation Concepts.* US Department of Transportation, Federal Highway Administration, Office of Research and Management, Washington D.C.

Andreu, J., Capilla, J. and Sanchís, E. (1996). AQUATOOL, a generalized decision-support system for water-resources planning and operational management. *Journal of Hydrology* **177**(3–4), 269–291.

Arrow, K., Solow, R., Portney, P.R., Leamer, E.E., Radner, R. and Schuman, H. (1993). Report of the NOAA panel on contingent valuation, January 15. *Federal Register* **58**(10), 4601–4614.

Baan, P. (1996). *Methods for Integrated Assessment and Evaluation of Environmental Measures* [in Dutch]. Report T2062, Delft Hydraulics, The Netherlands.

Barbier, E.B. (1994). Valuing environmental functions: Tropical wetlands. *Land Economics* **70**(2), 155–173.

Barton *et al.* (2006). *Using Belief Networks in Pollution Abatement Planning. Example from the Morsa Catchment, South Eastern Norway*. NIVA Report No.5213-2006, Norwegian Institute for Water Research.

Bateman, I.J., Brainard, J.S. and Lovett, A.A. (1995). Modelling woodland recreation demand using geographical information systems: A benefits transfer study. *Centre for Social and Economic Research on the Global Environment (CSERGE), University of East Anglia and University College London, England.*

Change Working Paper 95-06, Centre for Social and Economic Research on the Global Environment (CSERGE), University College London and University of East Anglia, Norwich, UK.

Bateman, I.J. and Willis, K.G. (Eds.) (1999). *Valuing environmental preferences. Theory and Practice of the Contingent Valuation Method in the US, EU and Developing Countries*. Oxford University Press, Oxford, UK.

Bateman, I.J., Carson, R.T., Day, B., Hanemann, M., Hanley, N., Hett, T., Jones-Lee, M., Loomes, G., Mourato, S., Ozdemiroglu, E., Pearce, D.W., Sugden, R. and Swanson, J. (2002). *Economic Valuation with Stated Preference Techniques: A Manual*. Edward Elgar, Cheltenham, UK.

Boeters, R. and Brouwer, R., *et al.* (2006). *Interregional comparison of the methodology to assess the cost-effectiveness of programs of measures in the International Scheldt District* [in Dutch]. International Scheldt Commission, Antwerpen, Belgium.

Bohm, P., and Russell, C.S. (1985). Comparative analysis of alternative policy instruments. In: Kneese, A.V. and Sweeney, J.L. (Eds.) *Handbook of Natural Resource and Energy Economics, Vol. I.* North-Holland, Amsterdam, The Netherlands.

Bovenberg, A.L., van den Broek, M.E.T. and Mulder, R.J. (1991). *Instrument choice in environmental policy* [in Dutch]. Discussion note 9102. Ministry of Economic Affairs, The Hague, The Netherlands.

Braat, L.C. and Lierop, W.F.J. (1987). *Integrated Economic-ecological Modelling.* North-Holland, Amsterdam, The Netherlands.

Brisson, I. and Pearce, D.W. (1995). Benefits transfer for disamenity from waste disposal. *Global Environmental Change.* Working Paper WM 95-06, Centre for Social and Economic Research on the Global Environment (CSERGE), University College London and University of East Anglia, Norwich, UK.

Brouwer, R. (2000). Environmental value transfer: State of the art and future prospects. *Ecological Economics* **32**, 137–152.

Brouwer, R., Turner, R.K. and Georgiou, S. (2003). Integrated assessment and sustainable water and wetland management. A review of concepts and methods. *Integrated Assessment* **3**(4), 171–183.

Brouwer, R. and van Ek, R. (2004). Integrated ecological, economic and social impact assessment of alternative flood protection measures in the Netherlands. *Ecological Economics* **50**(1–2), 1–21.

Brouwer, R. *et al.* (2004). *Assessment of environmental and resource costs in the Water Framework Directive*. Information sheet prepared by Drafting Group ECO2,

Common Implementation Strategy, Working Group 2B. RIZA working document 2004.203X, December 2004.

Brouwer, R., Schenau, S. and van der Veeren, R. (2005). Integrated river basin accounting and the European Water Framework Directive. *Statistical Journal of the United Nations Economic Commission for Europe* **22**(2), 111–131.

Brouwer, R. (2005). *Uncertainties in the economic analysis of the European Water Framework Directive*. IVM report E05-03. July 2005. Vrije Universiteit Amsterdam, The Netherlands.

Brouwer, R. and Hofkes, M.H. (Forthcoming). Integrated river basin modeling: An overview of key issues and their role in water policy management. *Ecological Economics Special Issue Integrated River Basin Modeling: Approaches and Key Issues.*

Cai, X. and McKinney, D.C. (1997). *A multiobjective model for negotiations in regional water resources allocation*. ASCE Water Resources Planning Management Division, 1997 Annual Conference. Houston, Texas.

Cai, X. and Wang, D. (2006). Calibrating holistic water resources-economic models. *Journal of Water Resources Planning and Management* **132**, 414–423.

Carson, R.T., Wright, J.L., Carson, N.J., Alberini, A. and Flores, N.E. (1995). *A bibliography of contingent valuation studies and papers*. Natural Resource Damage Assessment Inc., San Diego, CA.

Cave, R., Brouwer, R., Coombes, E., Hadley, D., Turner, R.K. and Lorenzoni, I. (2003). *Futures Scenarios and Cost-Effectiveness Analysis of Alternative Policy Options to Improve Water Quality in the Humber Catchment in the UK*. EUROCAT Project Report, CSERGE, UEA, Norwich, UK, p. 130.

Costanza, R., d'Arge, R., de Groot, R., Farber, S., Grasso, M., Hannon, B., Limburg, K., Naeem, S., O'Neill, R.V., Paruelo, JU., Raskin, R.G., Sutton, P. and van den Belt, M. (1997). The value of the world's ecosystem services and natural capital. *Nature* **387**, 253–260.

Dalhuisen, J.M., Florax, R.J.G.M., de Goort, H.L.F. and Nijkamp, P. (2003). Price and income elasticities of residential water demand: A meta-analysis. *Land Economics* **79**(2), 292–308.

Desvousges, W.H., Naughton, M.C. and Parsons, G.R. (1992). Benefit transfer: Conceptual problems in estimating water quality benefits using existing studies. *Water Resources Research* **28**(3), 675–683.

DETR (1998). *Policy Appraisal and the Environment: Policy Guidance*. DETR, London, UK.

Dudley, N.J. (1972). Irrigation planning, 4, Optimal interseasonal water allocation. *Water Resources Research* **8**(3), 586–594.

EEA (2005). *Market-based instruments for environmental policy in Europe*. EEA Technical report no. 8/2005. European Environment Agency, Copenhagen, Denmark.

EEA (2006). *Using the market for cost-effective environmental policy. Market-based instruments in Europe*. EEA Report no. 1/2006. European Environment Agency, Copenhagen, Denmark.

ERM, Stone & Webster (2004). *Assessing current levels of cost-recovery and incentive pricing*, DEFRA, May 2004.

FAO (2004). *Water charging in irrigated agriculture: an analysis of internal experience*. FAO Water Reports no. 27, Rome, Italy.

Foster, J. (Ed.) (1997). *Valuing Nature? Ethics, Economics and the Environment*. London, New York: Routledge.

Freeman III, A.M. (1993). *The measurement of environmental and resource values, Theory and methods*. Resources for the Future, Washington, D.C.

Gren, I.-M., Folke, C., Turner, R.K. and Bateman, I. (1994). Primary and secondary values of wetland ecosystems. *Environmental and Resource Economics* **4**, 55–74.

Hanley, N. and Spash, C.L. (1993). *Cost-benefit Analysis and the Environment*. Edward Elgar Publishing Limited, England.

Holland, M., Hunt, A., Hurley, F., Navrud, S., Watkiss, P. (2005). *Methodology for the Cost-Benefit Analysis for CAFÉ: Volume 1: Overview of Methodology*. Report to DG Environment, European Commission, Brussels, Belgium.

Johansson, P.O. (1987). *The economic theory and measurement of environmental benefits*. Cambridge University Press, Cambridge, UK.

Kask, S.B. and Shogren, J.F. (1994). Benefit transfer protocol for long-term health risk valuation: a case of surface water contamination. *Water Resources Research* **30**, 2813–2823.

Larson, J.S., Adamus, P.R. and Clairin Jr, E.J. (1989). *Functional Assessment of Freshwater Wetlands: A Manual and Training Outline*. WWF and the Environmental Institute, University of Mass. Amerst, USA.

Laurans, Y. (2004). An example of cost recovery assessment in Seine-Normandie. In: Brouwer, R. and Strosser, P. (Eds.) *Environmental and Resource Costs and the Water Framework Directive. An Overview of European Practices*. Workshop Proceedings. Amsterdam, 26th March 2004. RIZA working paper 2004.112x. RIZA, Lelystad, The Netherlands.

Loomis, J.B., Roach, B., Ward, F. and Ready, R. (1995). Testing transferability of recreation demand models across regions: a study of corps of engineer reservoirs. *Water Resources Research* **31**, 721–730.

Luken, R.A., Johnson, F.R. and Kibler, V. (1992). Benefits and costs of pulp and paper effluent controls under the Clean Water Act. *Water Resources Research* **28**, 665–674.

Maltby, E., Hogan, D.V. and McInnes, R.J. (Eds.) (1995). *Functional Analysis of European Wetland Ecosystems*. Final Report – Phase One. EC DGXII STEP Project. CT90-0084. Wetland Ecosystems Research Group. University of London, UK.

McKinney, D.C. and Cai, X. (1997). Multiobjective optimization model for water allocation in the Aral Sea basin. Second American Institute of Hydrology (AIH) and Tashkent Institute of Engineers for Irrigation (IHE) *Conjunct Conference on the Aral Sea Basin Water resources Problem*. Tashkent, Uzbekistan.model for negotiations in regional water resources allocation. ASCE Water Resources Planning Management Division, 1997 annual conference, Houston, Texas.

McKinney, D., Cai, X., Rosegrant, M.W., Ringler, C. and Scott, C.A. (1999). *Modeling Water Resources Management at the Basin Level: Review and Future Directions*. Colombo, International Water management Institute, SWIM Paper 6, pp. 71.

Mitchell, R.C. and Carson, R.T. (1989). *Using surveys to value public goods: The contingent valuation method*. Resources for the Future. Washington D.C.

Nichols, D.S. (1983). Capacity of natural wetlands to remove nutrients from wastewater. *Journal of Water Pollution Control* **55**, 495–505.

OECD (1997). *Evaluating Economic Instruments for Environmental Policy*. Organisation for Economic Co-operation and Development, Paris, France.

OECD (1998). *Agricultural Water Pricing in OECD Countries*. Working party on Economic and Environmental Policy Integration. ENV/EPOC/GEEI(98)11/FINAL. Organisation for Economic Co-operation and Development, Paris, France.

OECD (1999). *The Price of Water. Trends in OECD Countries*. Organisation for Economic Co-operation and Development, Paris, France.

Oosterhuis, F. (2006). *Experiences with tradable permits in water quality policy in the United States* [in Dutch]. Water and Economic Modeling for Policy Analysis (WEMPA) working paper W-06/29. Institute for Environmental Studies, Vrije Universiteit Amsterdam, The Netherlands.

Pearce, D.W. and Turner, R.K. (1990). *Economics of Natural Resources and the Environment*. Harvester Wheatsheaf, UK.

Pearce, D.W., Whittington, D. and Georgiou, S. (1994). *Project and Policy Appraisal: Integrating Economics and Environment*. Organisation for Economic Co-operation and Development, Paris, France.

Pretty, J.N. *et al.* (2000). An assessment of the total external costs of UK agriculture. *Agricultural Systems* **65**, 113–136.

Randall, A. (1983). The problem of market failure. *Natural Resources Journal* **23**, 131–148.

Randall, A. (1988). A camel is a camel is .... In: Peterson, G.L., Driver, B.L. and Gregory, R. (Eds.). *Amenity Resource Valuation: Integrating Economics with Other Disciplines*. Venture Publishing Company.

Renzetti, S. and Kushner, J. (2004). Full cost accounting for water supply and sewage treatment: Concepts and application. *Canadian Water Resources Journal* **29**(1), 13–23.

Rosegrant, M., Ringler, C., McKinney, D.C., Cai, X., Keller, A., and Donoso, G. (2000). Integrated economic-hydrologic water modelling at the basin scale: the Maipo River Basin. *Journal of Agricultural Economics* **24**(1): 33–46.

Sagoff, M. (1988). Some problems with environmental economics. *Environmental Ethics* **10**: 55–74.

Schaafsma, M. and Brouwer, R. (2006). *Overview of existing guidelines and manuals for the economic valuation of environmental costs and benefits*. AquaMoney Report (www.aquamoney.org), Amsterdam, IVM, The Netherlands.

Stoneham, G., Chaudhri, V., Ha, A. and Strappazzon, L. (2003). Auctions for conservation contracts: An empirical examination of Victoria's Bush Tender trial. *The Australian Journal of Agricultural and Resource Economics* **47**(4): 477–500.

Task Force on Water Satellite Accounts (2002). *Recent Developments in Water Policy and Statistics*. Doc. Water TF/02/3. EUROSTAT, Luxembourg.

Turner, R.K. (1993). Sustainability: Principles and Practice. In Turner, R.K. (eds.) *Sustainable Environmental Economics and Management: Principles and Practice*. Belhaven Press, London, UK.

Turner, R.K., Georgiou, S., Clarke, R. and Brouwer, R. (2004). *Economic Valuation of Water Resources in Agriculture. From the Sectoral to a Functional Perspective of Natural Resource Management*. FAO Water Report 27, FAO, Rome, Italy.

UN (2006). *Integrated environmental and economic accounting*. United Nations.

van der Bolt, F., van den Bosch, R., Brock, T., Hellegers, P., Kwakernaak, C., Leenders, D., Schoumans, O. and Verdonschot, P. (2003). *Aquarein; Consequences of the European Water Framework Directive for agriculture, nature, recreation and fishery* [in Dutch]. Alterra-report 835. Wageningen University en Research Centre, The Netherlands.

Velázquez, E. (2005). An input–output model of water consumption: Analysing intersectoral water relationships in Andalusia. *Ecological Economics* **56**: 226–240.

VEWIN (2002). *Water Supply Statistics* 2002. VEWIN The Netherlands Waterworks Association, Rijswijk, The Netherlands.

WATECO (2002). *Economics and the Environment. The Implementation Challenge of the Water Framework Directive.* A Guidance Document.

Young, R.A. (2005). *Determining the Economic Value of Water. Concepts and Methods.* Resources for the Future, Washington DC, USA.

Zhang, Z. and Folmer, H. (1995). *Economic Approaches to Cost Estimates for the Control of Carbon Dioxide Emissions.* Wageningen Economic Papers no. 1995-2. Faculty of Economics. Wageningen Agricultural University, The Netherlands.

7. ANNEX: REVIEW OF SELECTED INTEGRATED HYDRO-ECONOMIC MODELS

Table 1 provides an overview of models reviewed in this annex[5]. The selected models are by no means representative of the hydro-economic modelling work carried out in the countries where the models have been developed. The selection is based on the aim to include a broad range of possible approaches for different water resources of relevance to the WFD. Several of the models have also been presented during the international workshop on Integrated River Basin Modelling and the EU Water Framework Directive 17–18 November 2005 at the Vrije Universiteit Amsterdam (www.ivm.falw.vu.nl/watereconomics).

Broadly speaking the models address either water allocation between competing uses or water quality issues. In accordance with the conceptual hydro-economic model description in Figure 4.2, models are either classified as single or multi-sector water use models. Economics is integrated in the water allocation or water quality models either partly as a 'post-processing module' or fully. None of the reviewed models can be classified as fully integrated water quantity–quality models. Where both quality and quantity modules are available, water quality models are simple mass-balance models, which post-process water quality based on hydrological simulation (MIKE BASIN, AQUATOOL, Water StrategyMan DSS). The models reviewed here that have water quality modules

[5]A review of additional hydro-economic decision-support systems can be found in ProGea (2004).

focus on nutrients and algal biomass (eutrophication), BOD, OD or faecal coliforms in view of the fact that the underlying science for the behaviour of these pollutants is relatively well established (MIKE, AQUATOOL, Water StrategyMan DSS water quality simulators, Bayesian network using eutrophication models in Norway).

Very few of the models presented here include a fully integrated economic optimisation model with hydrological simulation, possibly due to the computational limitations of dealing with multiple competing uses and simultaneous flow and quality constraints. In general, we find that only single sector models are fully integrated hydro-economic models. Powel Integrated Control Centre (ICC) is an illustration of a model focusing on a single water user (hydropower), which allows for seamless integration of a market simulator with a hydrological water allocation model. The DSIRR model is another example applied to the irrigated agricultural sector.

Table 1 Hydro-economic model overview

Quantity/ quality	Economic water use models	Hydrological/ biochemical models with economic post processing*	Fully integrated hydroeconomic model	Meta-models
Water allocation	Single sector	None reviewed	– Powel Integrated Control Centre (ICC) (hydropower) – Decision-support for irrigation (DSIRR)	– Hugin Bayesian networks
	Multi-sector	– MIKE Basin with Excel/VBA optimisation – AQUATOOL – Water StrategyMan DSS	None reviewed	

(*continued*)

Table 1 *Continued*

Quantity/ quality	Economic water use models	Hydrological/ biochemical models with economic post processing*	Fully integrated hydroeconomic model	Meta-models
Water quality	Single sector	None reviewed	None reviewed	
	Multi-sector	– MIKE Basin with Water Quality simulator (nutrients) – Ad hoc "model chains" in PIREN-Seine modelling toolbox (nutrients, algal blooms) – AQUATOOL water quality simulator – Water StrategyMan DSS water quality simulator	None reviewed	– Hugin Bayesian networks

Ad hoc linking together of models from a toolbox is another approach, which may be more commonplace in the WFD implementation across Europe in view of the fact that calibration and validation as in the case of new model development is generally not required. This approach is used in France for scenario and cost-effectiveness analysis (PIREN-Seine modelling toolbox). Whereas these ad hoc model chains are usually deterministic in nature, Bayesian networks (BN) focus explicitly on uncertainty in the modelling chain, using Monte Carlo simulation of probability distributions for the uncertain variables in the underlying models. BN link probability distributions together in a meta-model, making it possible to conduct decision and information analysis under uncertainty.

Based on the review in this annex, Table 2 summarises how relevant and useful the reviewed models are with regards to the different aspects in the economic analysis underpinning the WFD. Any catchment that has an established and functioning hydro-economic model should have – as a rule – adequate data for the WFD reporting requirements regarding the economic characterisation of river basins for the sectors that are included in the model. Apart from that we find that no single model covers all economic elements. Most models cover one specific field, e.g., dynamic optimisation and the study of price effects, while fully integrated models are usually limited to a single sector. Although Baysian Belief Networks are suited for decision-analysis regarding the ranking of measures under uncertainty, as a more generic tool it lacks the specific functionalities required for for example a full cost analysis. The only tool that basically has most of the required functionalities to some extent for multiple water users of a river basin is Water Strategy Man. This model was designed with the WFD in mind though. The model's flexible application on several WFD related issues comes at the price of accuracy and pre-processing/ assumptions about input data.

Table 2 Relevance of the selected integrated hydro-economic models for the economic analysis in the WFD

	ICC	DSIRR	MIKE Basin	AQUATOOL	Water StrategyMan DSS	Baysian belief networks
Scenario analysis						
CEA ranking of measures	Allocation measures	Water allocation, land use, agricultural practices	Allocation measures	Allocation measures	Allocation measures	Generic decision analysis
CBA ranking of measures	Optimisation water allocation water users	Optimisation water allocation farms				Generic decision analysis
Resource costs						
Environmental damage costs			Resource costs of environmental streamflows	Resource costs of environmental streamflows	Based on administrative charges	
Price incentive effects						
MCA ranking of measures					Multiple scenario indicators	

Note: Yes
Partial
No

A. Optimising hydropower generation with Powel Integrated Control Centre (ICC) (Samkjøringsmodellen & Vansimptap)

Table A.1 Powel Integrated Control Centre (Market Analyzer and Generation Optimizer)

Kind of model (categorization)	Stochastic dynamic programming. Modular
DPSIR description	Drivers (power supply contracts and demand); Pressure (hydropower regulation of water bodies); State (reservoir levels, river flow); Impact (value of water in reservoir); Response (optimal generation strategies across generation area)
Water use functions/emission sources included	Hydropower generation
Substances included	None
Comparison with emission and /or water quality standards	Used to evaluate alernative definitions of good ecological potential in heavily modified water bodies
Priority substances of the WFD	None
Ease of use	Commercial software. Windows based. Modular. Advanced user requirements for model calibration.
Expected accuracy of results	Weekly temporal resolution, independent of available resolution of hydrological data. Weekly resolution is (probably) excessive relative to resolution of ecological requirements of indicator species (seasonal?)
Temporal scale and resolution of model and results (seasonality?)	The Sim model is optimised for weekly intervals, based on previous weeks tapping distribution and a weekly description of supply and demand (power contracts). Within a single week contracts may specify price tranches in which case Sim provides optimal generation for each tranche.
Spatial scale and resolution of model and results	Flexible. Hydrological network defined in Sim may reflect sub-catchment, catchment or several catchments tied together by common transmission lines. Resolution at level of individual power plant, reservoir, river reach.
Down-scaling or up-scaling of results	Flexible upscaling by coupling several power generation areas together in Generation Optimizer. Downscaling below/within water body not possible (e.g. analysis of effects of hydromorphological measures)
Sensitivity and scenario analysis capabilities	Using alternative specification of model structure (adding, subtracting demand segments, supply/ production units); alternative model parameter definition (e.g., changing tapping restrictions)

(*continued*)

Table A.1 *Continued*

Uncertainty/simulation analysis capabilities	Stochastic dynamic programming takes into account historic variation of inflow to hdyropower system. Probabilistic value of water analysis. Probablistic approach allows user to select requisite levels of confidence.
Economic demand and supply functions	Demand defined through fixed supply and variable supply contracts which enter the market at pre-defined prices. Supply defined by power plant capacity and reservoir tapping restrictions.
Economic optimisation functionality (cost-effectiveness, efficiency analysis, cost-benefit anlaysis)	Efficiency analysis by evaluating market clearing power price. Calculation of consumer and producer surplus possible.
Model transferability	Generic modular modelling tool. Transferable across catchments and power markets. Used country-wide by Norwegian regulator and by most power companies.
Procedure of linking water with economic issues	Stochastic Dynamic Programming. Hydrological and power generation network simplified to a single aggregated reservoir and power station in order to calculate and optimal production strategy – optimal solution then allocated to the different reservoirs based on a rules-based tapping strategy
Assessment of cost recovery, including environmental and resource costs	Can be used to evaluate the opportunity costs to hydropower of the streamflow required to meet good ecological status or good ecological potential.

Introduction

The Powel Integrated Control Centre is a suite of models designed for short and long term optimisation of hydropower generation, coupling physical models of the hydrological and hydropower system with models of the eletricity market. In this screening we review the Powel Generation Optimizer and Powel Sim. As implemented in Norway the models go by the names of "Samkjøringsmodellen" and "Vansimtap" respectively and are i.a. used by the Norwegian Directorate for Energy and Water Management (NVE) to evaluate costs of definitions "ecological potential" under the WFD for hydropower producers and consumers in Norway. The full suite of models is marketed by Powel ASA as the Powel Integrated Control Centre (ICC) (http://www.powel.no/) and includes:

Powel Sim – Consequence analysis, Restriction- and Water Management
Powel Shop – Short Term Hydro Generation
Powel Market Analyzer – Long Term Market

Powel Generation Optimizer – Long Term Water Value
Powel Inflow – Inflow Forecaster and Analyzer
Powel Autopilot – Automatic ICC generation systems
Powel ICC Web – ICC Intranet Reporter

Powel Generation Optimizer is a one-area power-market simulator for optimizing long- and medium term capitalization of hydropower resources and contract access. The optimal solution consider uncertain inflow, market prices and delivery obligations. It is a tool for managing and evaluating hydropower generation system in complex watercourses. The model can generate water-values and boundary conditions for the short-term optimizer Powel Shop[6]. In Norway Powel Generation Optimizer is used to set boundary conditions for the program Powel Sim which calculates the consequences of generation plans and inflow forecasts in a hydropower system, on short- or medium-term. Powel Sim is used to increase the efficiency of hydropower producing companies by allowing them to look into how different generation plans follow planned progress for the reservoir in different running situations, or to find a running schedule that makes the most of the resources in short-term generation planning[7].

Sim is being used to evaluate hydropower loss relative to the optimal solution of alternative definitions of "ecological potential" in reservoirs and water courses within a single 'power generation area'. Power generation areas can be defined flexibly in extent and number (e.g., power stations within hydrological networks can be aggregated within any grid definition). Generation Optimizer is used to find the medium to long term value of water in reservoirs on the margin by equating the price of demand side power contracts with supply side production possibilities in a wider power market. Prices in this market set boundary conditions for optimal production within any power generation area.

Interaction water economics

The full modular structure of Powel ICC is shown in Figure A.1. For definition and evaluation of 'good ecological potential' in Norway only parts of the modular system are being used. Generation Optimizer calculates medium to long-term water values based on the demand and supply curves for the generating area being studied ('restriction management'), producing a set of iso-price curves. Sim is used to evaluate optimal hydropower production given the iso-price curves determined in Generation Optimizer ('hydro-power

[6]http://www.powel.no/pictures/ProductFactSheetICCGenerationOptimizer.pdf
[7]http://www.powel.no/pictures/ProductFactSheetICCSim.pdf

simulator'). Inflow is given as historical time series to Sim (the Inflow Forecaster is not used here).

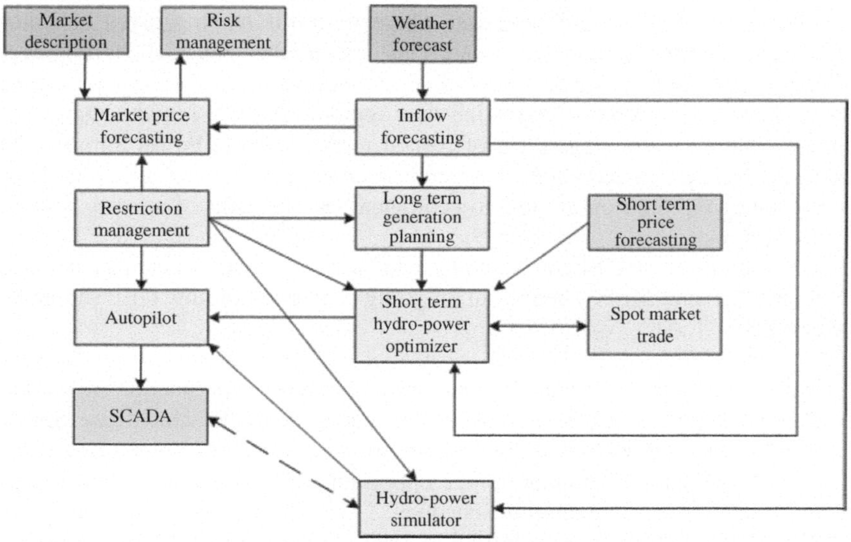

Figure A.1 Powel Integrated Control Centre modules. (Source: http://www.powel.no)

Given the need to optimize generation with variability in inflow, Sim uses Stochastic Dynamic Programming which requires that the hydrological and power generation network is simplified to a single aggregated reservoir and power station in order to calculate and optimal production strategy. The optimal solution is then allocated to the different reservoirs based on a rules-based tapping strategy. Buffer reservoirs are tapped based on reservoir steering curves in order to cover predefined power effect requirements for a specific power stations, only deviating from these when system-wide restrictions are reached. Regulated reservoirs are tapped based on a predefined set of restrictions, which include ecological and user requirements.

Scenarios and policy measures

'Good ecological potential' (GEP) for 'heavily modified water bodies' (HMWB) such as regulated reservoirs and rivers is currently being evaluated in pilot watersheds in Norway. Achievement of GEP in HMWB will depend on a combination of ecological tapping restrictions and physical measures within the water bodies. Whereas 'poor ecological potential' (PEP) may be given when power generation is optimised regardless of tapping restrictions, 'maximum ecological potential' (MEP) is given where reservoir tapping gives indicator organisms optimal ecological conditions given the physical hydropower infrastructure and its hydromorphological modifications of water bodies. Powel Sim can be used to determine flow regimes at PEP. Implementation of all available measures (regardless of cost) determines MEP. Powel Sim may be used to determine maximum opportunity costs (relative to PEP) of e.g., completely adapting flow regime to ecological requirements of species without regard for costs.

The definition of GEP in Norway is located somewhere between the two extremes PEP and MEP. There is an on-going discussion of how GEP should be determined[8].

(i) a definition of the local sustainability of indicator species may be used to determine flow and morphological requirements, in which case Sim is used to compute resulting opportunity costs to power generation (of the flow/tapping restrictions); here opportunity cost is defined relative to an unconstrained generation schedule
(ii) the regulator fixes a definition of disproportionate cost for any particular power plant (e.g., 5% of current gross income at each power station) and Sim is used to calculate resulting tapping strategies/flow for the associated reservoirs; here opportunity cost is defined relative to the current concession regime. This may or may not coincide with a flow regime that sustains target species.

Generation optimizer can also be used evaluate external costs of e.g., replacing hydro-power with carbon based power sources.

[8]Disproportionate cost is dependent on what is used as a baseline (0 scenario): power producers may argue that this is an unconstrained regime, whereas it is in the interest of the regulator to refer to the strictest existing regime from the point of view of ecological restrictions.

B. MIKE Basin & Water Quality Simulator (WQ), including Optimisation using Excel/Visual Basic (VBA) macros

Table B.1 MIKE BASIN, INCL. WQ & VBA

Kind of model (categorization)	Object-oriented hydrological simulation and optimisation model
DPSIR description	Pressure (water allocation between uses; pollution loading); state (water availability, water quality)
Water use functions/emission sources included	Water user objects: themes for irrigation, industrial/ municipal water supply, reservoirs etc. Water resource objects: node, branch, catchment properties, groundwater
Substances included	water run-off and allocation platform with possibility for water quality simulations (add-in). Water quality simulator: first-order decay functions for NH4/NO3, DO/BOD, COD, E-coli, Ptot and wild card substances. Load calculator for non-point pollution including NH4, NO3, E-coli and Ptot.
Comparison with emission and/or water quality standards	No
Priority substances of the WFD	No
Ease of use	Requires ArcGIS or ArcView; GIS knowledge and platform. Ease of establishing water node-branch network. Handles hydrological timeseries directly in a "time series editor".
Expected accuracy of results	Depends on resolution of hydrological input data.
Temporal scale and resolution of model and results (seasonality?)	Daily or even hourly resolution depending on available hydrological data.
Spatial scale and resolution of model and results	Catchment level. Resolution only limited by availability of GIS data on flow network. Size limits on flow network?
Down-scaling or up-scaling of results	Flow network can be visually scaled according to desired resoultion in GIS interface.
Sensitivity and scenario analysis capabilities	Programmed in VBA macros
Uncertainty/simulation analysis capabilities	None in the sense of Monte Carlo simulation. However, dynamic optimisation provides an evaluation of temporal variation which for any given time period can be interpreted as uncertainty. No tool to evaluate probabilities of threshold value exceedance per time period.

(*continued*)

Table B.1 *Continued*

Economic demand and supply functions	No. Water demand functions may be specified outside MIKE Basin in Excel. Economic optimisation of water allocation is done in Excel Solver. VBA macros are used to run either single simulations or time-step-by-time step analyses
Economic optimisation functionality (cost-effectiveness, efficiency analysis, cost-benefit analysis)	No, but linear and non-linear optimisation can be carrier out in Excel Solver.
Model transferability	Transferable generic modelling tool
Procedure of linking water with economic issues	Post-processing using VBA and Excel Solver. MIKE Basin carries out optimal water allocation between users based on water demand functions specified in Excel.
Assessment of cost recovery, including environmental and resource costs	Resource costs. Opportunity costs to any water user can be calculated based on a reallocation of water between users. Streamflow that satisfies GES/GEP can be defined as a water user.

Introduction

MIKE Basin[9] is a tool for analysing catchment-wide water allocation between competing uses. It is one of a suite of water resource management tools sold by the Danish Hydrological Institute (DHI). It is the only catchment water resource management tool that allows model set up in ArcGIS, easing integration with other ESRI GIS products. MIKE Basin includes a pollution load calculator for summing loads for point and non-point sources for sub-catchments. Water quality is calculated using simple steady-state equations assuming perfect mixing in water bodies. Water allocation is rules-based – economic optimisation is not supported within MIKE Basin itself, but can be handled using a combination of VBA macros and Excel Solver. Although developed in Denmark the model has found wide application in many countries due to its generic structure and ESRI platform.

Interaction water economics

MIKE BASIN is both a simulation and an optimization model, with a built-in nonlinear program solver (based on a sequential quadratic program algorithm). MIKE BASIN can optimize with respect to any objective, including water quality targets. Given that economic data is not supported within MIKE Basin

[9]http://www.dhigroup.com/Software/WaterResources/MIKEBASIN.aspx

optimization is related to physical goals and constraints (e.g., allocation rules) connected to each water body. DHI have chosen instead to support communication between MIKE Basin and Excel, the latter being a common everyday tool for economic analysis. Visual Basic macros must be programmed for the specific water use optimisation at hand, inputing demand and supply data in Excel and using Excel Solver for the economic optimisation. Simulation results from MIKE Basin are imported directly to the spreadsheet, where one can then take advantage of Excel's suite of economic functions (discounting, investment analysis, etc.), standard graphs, and the Solver tool for optimization. Excel data can also influence the simulation itself, for example in analysing effects on demands by pricing water according to availability. Simulations can be run one time step at a time, or across a multi-period allocation problem.

Scenarios and policy measures

Despite its generic features, MIKE Basin is primarily designed to handle water allocation problems, given that most of its input data is related to hydrology. As a simulation tool it can be used to evaluate scenarios of e.g., changes in catchment inflow and water use. Standard rules for reservoir operation can be chosen from drop down tables, or specific rules programmed in VBA for particular reservoirs of interest. Optimisation on water flow, pollution loads or steady-state concentrations can be handled within MIKE Basin for any spatial disaggregation chosen by the user, including single reservoirs, sub-catchments or whole catchments. Using Excel first order effects of price changes can be modelled (e.g., water user fees, pollution emissions charges). Any water use demand function that can be specified in Excel can in principle be used for policy analysis with MIKE Basin. Optimisations are partial equilibrium and static in the sense that prices and demand functions are fixed for the water allocation period simulated in MIKE Basin.

C. Eutrobayes – Bayesian belief networks of economic analysis of nutrient mitigation measures

Table C.1 Bayesian network/influence diagram

Kind of model (categorization)	Bayesian belief networks, influence diagrams
DPSIR description	Any uncertain causal links in the DPSIR model are supported. In Barton *et al.* (2006), Drivers (agricultural area, population size), Pressures (nutrient loading), States (nutrient and algal concentrations, % toxic blue-green algae), Impacts (costs of remediation measures, benefits of swimmable lake water), Responses (alternative nutrient mitigation measures).
Water use functions/ emission sources included	Eutrophication and toxic algal blooms Emissions from agricultural run-off, non-point and point source municipal sewage
Substances included	Phospohorous
Comparison with emission and/or water quality standards	Yes. Probability of non-compliance with existing Norwegian standards for water user suitability (total phosphorous, Chlorophyl A, turbidity), as well as selected indicators of "good ecological status" in the Water Framework Directive (Chlorophyl A, % blue-green algae).
Priority substances of the WFD	None.
Ease of use	Graphical user interface for defining network structure. Windows menus for entering conditional probabilities into network nodes. Advanced user knowledge required for importing conditional probability tables from underlying models (using VBA macros). Slow run times/crashing of networks with excessive causal linkages between nodes.
Expected accuracy of results	Discretisation of probability tables must be done 'manually', being time consuming. Discretisation assumptions introduce an additional source of uncertainty into BNs.
Temporal scale and resolution of model and results (seasonality?)	Annual remediation costs are evaluated against effects on eutrophication during the bathing season (June–September). No inter-annual effects modelled.
Spatial scale and resolution of model and results	Network implemented for the Morsa catchment. Aggregation of nutrient sources can be downscaled to study effects of sub-catchment level mitigation efforts. Scaling limitations depend on domain of underlying models. Run-off model (USLE) cannot be downscaled to plot level as it has difficulty in dealing with different forms of erosion. Scaling to catchments with multiple lakes/reservoirs requires specifying nutrient retention in river reaches and intermediate water bodies. Scaling/ transferring willingness to pay data to other water bodies possibly limited by site specific recreational users.

(*continued*)

Table C.1 *Continued*

Down-scaling or up-scaling of results	Not flexible for any given network. But nesting of sub-catchments in object oriented Bayesian networks (OOBNs) makes adjusting networks relatively easy (cut-and-paste)
Sensitivity and scenario analysis capabilities	Conducted on conditional probabilities in network nodes and on model structure (presence absence of nodes and causal linkages between them). Sensitivity analysis on parameters of underlying models possible for those models that have Monte Carlo simulation routines (probabilistic model outcome must be simulated conditional on alternative parameter values; parameters of interest are then made explicit in the BN as unconditional marginal probability nodes).
Uncertainty/simulation analysis capabilities	No simulation capabilities. BNs rely on MC simulation/ uncertainty analysis capabilities of underlying models to produce conditional probability tables.
Economic demand and supply functions	Modelled as response surfaces in the form of conditional probability tables (a regression function can be modelled as a network)
Economic optimisation functionality (cost-effectiveness, efficiency analysis, cost-benefit analysis)	None, in terms of constrained optimisation. BNs can be used for ranking of alternative courses of action under uncertainty based on their expected utility (cost-effectiveness and cost-benefit analysis). Information value analysis is supported.
Model transferability	Transferability of problem structure: Bayesian networks should be transferable where causal problem structure is the same (i.e., when causes and effects of eutrophication are the same between catchments). Transferability of response surfaces: conditional probabilities are probably less transferable between sites (i.e., when lake typologies or water user population characteristics vary)
Procedure of linking water with economic issues	Bayesian belief networks are a generic approach to decision analysis where cause-effect linkages are described in terms of relationships between probability distributions (conditional probability distributions). See Figure C.1.
Assessment of cost recovery, including environmental and resource costs	No

Introduction

Bayesian belief networks are a generic tool for integrating empirical relationships, model simulations and expert opinion in decision analysis under uncertainty. Bayesian belief networks are seeing increasing application to handling of uncertainty in environmental and natural resource management (Varis and Kuikka, 1999). Several applications to water quality/eutrophication have appeared in recent literature (Ames *et al.*, 2005, Borsuk *et al.*, 2004) and the Bayesian approach has been proposed as a way of dealing with uncertainty in economic analysis under the Water Framework Directive (Barton *et al.*, 2006). The EU project MERIT[10] has also been developing Bayesian networks for a number of water management problems including domestic water demand, pesticide pollution impacts on groundwater, competing demands of irrigation and hydropower, and water competition between domestic, environmental and agricultural sectors. A number of commercial and public software applications are available for implementing Bayesian networks – here we refer to functionality of Hugin Expert[11].

Interaction water economics

Hugin is a generic tool for decision analysis in any modelling domain where causal relationships observed in data, modeled analytically or defined by expert opinion can be implemented. As such the integration between economics and water related sciences in any particular application depends more on the availability of empirical data, analytical models and relevant and willing expertise than on properties of Bayesian networks (BNs) in e.g., Hugin. Nevertheless, some general constraints on implementing models in Hugin also apply to the integration of economics (i) for Bayesian statistics algorithms to work, BNs must be acyclical, meaning that feedback loops in a causality chain represented by a network are not permitted. This in turn means that dynamic modelling within Hugin is impossible and that time series beyond a few periods are difficult to model explicitely (a separate network must be specified for every time period of interest). This in turn means that discounting is not easily carried out within BNs. Structural relationships in data, models or opinion are summarised in 'response surfaces' in the form of conditional probability tables. A BN functions as a kind of meta-model linking these response surfaces together. Time series and discounting are therefore best handled in models that provide these response surfaces[12]. (ii) Linear and non-linear optimisation in the economic sense of maximising an objective function subject to constraints cannot be conducted in a BN. The form of optimisation BNs are designed for is

---

[10] http://www.geus.dk/program-areas/water/denmark/rapporter/merit_aug_04-uk.htm

[11] http://www.hugin.com/

[12] For example, a dynamic water quality model may simulate daily algal concentrations in a lake used for bathing. Daily concentrations are summarised as probabilities of different concentration ranges within the bathing season. Willingness to pay for bathing water quality is calculated in Excel for different discount rates and confidence levels and results are summarised in a conditional probability table in a BN.

focused on evaluating what evidence is required of a response in order for some cause to be true, given an uncertain response surface (likelihood). Knowing the cost of obtaining such evidence and the costs and benefits of the response, information value can be determined and an efficient level of evidence gathering can be determined.

Scenarios and policy measures

Bayesian networks are a platform for decision analysis. In Hugin, Influence Diagrams are Bayesian Networks (linked conditional probability nodes) augmented with decision nodes and utility nodes (Figure C.1).

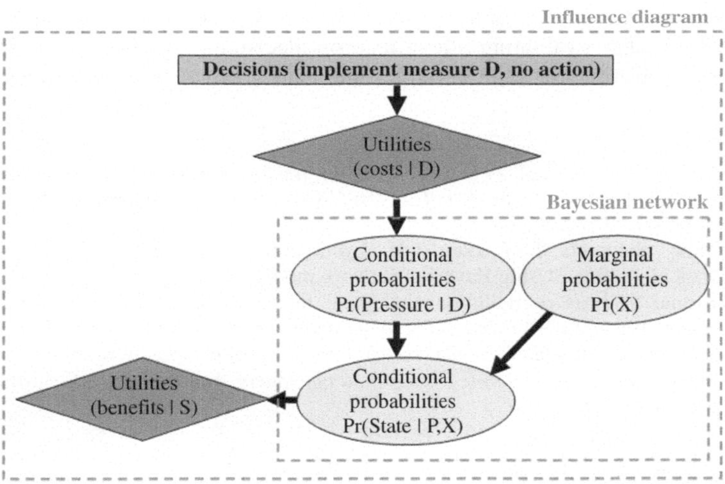

Figure C.1 Baysian networks and influence diagrams. (Source: Barton *et al.* (2006))

Utility nodes define uncertain costs and benefits of alternative actions. Any policy measure that can be formulated as alternative course of action – in its simplest form 'action' versus 'no action' – each with uncertain costs and benefits can be analysed in Hugin. In a Norwegian a case study (Barton *et al.,* 2006a), Influence Diagrams were used to evaluate a cost-efficient ranking of nutrient abatement measures under uncertainty, and to evaluate the economic risk of non-compliance with 'good water status' (probability of non-compliance × foregone benefits). This was possible in the selected catchment (Morsa, South-Eastern

Norway) because of available validated water quality models, and available studies of abatement costs of alternatives and willingness to pay to for different water uses.

Bayesian networks and influence diagrams distinguish themselves from analytical models in economic and natural science because they place greater focus on evaluating problem structure uncertainty than on parameter uncertainty. Scenario analysis in the sense of adjusting many underlying model parameters at once is not easy in a Bayesian network because it is meta-model in which the parameters of underlying models/response surfaces are not explicit. However, Bayesian networks make it easy to evaluate the effect of alternative prior beliefs about the strength and form of the causal relationships between the state variables of interest in a water body (e.g., the quality components that make up 'good ecological status'). Furthermore, the definition of scenario analysis can be extended to also evaluating the effect on decisions of alternative network structures with different state variables and causal linkages between them.

References

Ames, D.P. and Nielson, B.T. *et al.* (2005). Using Bayesian networks to model watershed management decisions: an East Canyon Creek case study. *Journal of Hydroinformatics* **7**(4): 267–282.

Barton, D.N. , Saloranta, T., Bakken, T.H., Lyche Solheim, A., Moe, J., Selvik, J.R. and Vagsta, N. (2006a). Using Bayesian network models to incorporate uncertainty in the economic analysis of pollution abatement measures under the water framework directive. *Water Science and Techology: Water Supply* **5**(6): 95–104.

Barton, D.N. and Saloranta, T. *et al.* (2006b). *Using Bayesian belief networks in pollution abatement planning. Example from Morsa catchment, South Eastern Norway.* NIVA Report SNO 5213-2006.

Borsuk, M.E., Stow, C.A. and Reckhow, K.H. (2004). A Bayesian network of eutrophication models for synthesis, prediction, and uncertainty analysis. *Ecological Modelling* **173**, 219–239.

Varis, O. and Kuikka, S. (1999). Learning Bayesian decision analysis by doing environmental and natural resources management. *Ecological Modelling* **119**, 177–195.

D. AQUATOOL

Table D.1 AQUATOOL

Kind of model (categorization)	Object-oriented hydrological simulation and optimisation model with an integrated post processing module for economic analysis
DPSIR description	Drivers (agricultural practices, urban and industrial activities), Pressures (changes in flow, pollution loading), State (water body levels, river flow, water body quality)
Water use functions/emission sources included	Water user objects: irrigation, industrial/municipal water supply, reservoirs etc. Water resource objects: node, branch, catchment properties, groundwater
Substances included	Water quality indices (basic water chemistry only?)
Comparison with emission and/or water quality standards	No
Priority substances of the WFD	No
Ease of use	Windows based graphical user interface for model set up. Considerable user experience required for hydrological model calibration and validation.
Expected accuracy of results	Depends on accuracy of hydrological input data although constrained to analysis of monthly demands
Temporal scale and resolution of model and results (seasonality?)	Monthly time scale, any geographical resolution.
Spatial scale and resolution of model and results	Flexible spatial scale. Resolution at water body scale (e.g., reservoirs modelled as a single object).
Down-scaling or up-scaling of results	Upscaling to annual and multi-annual totals.
Sensitivity analysis capabilities	No menu/automated options? Monthly node characteristics (operating restrictions, water demand functions etc.) may be modified manually in graphical user interface.
Scenario analysis capabilities	Using alternative hydrological inflow scenarios in input data files.
Uncertainty/simulation analysis capabilities	Simrisk module. 2–5 year risk analysis of supply failures for any node or group of nodes using stochastic inflow.
Economic demand and supply functions	Economic post processing module. Monthly water demand functions. Supply simulated in AQUATOOL based on operating restrictions.

(*continued*)

Table D.1 *Continued*

Economic optimisation functionality (cost-effectiveness, efficiency analysis, cost-benefit analysis)	Cost minimisation (scarcity cost and operating costs) subject to mass balance in nodes and reservoirs, capacity constraints, minimum flows, mass balance and flows in aquifers. Shadow prices of balance constraints in nodes of optimisation model provide marginal resource opportunity cost time series at specific location in the system.
Model transferability	Modular generic structure transferable. Management at monthly and basin scale (i.e. not suitable for daily hydropower plant operation optimisation)
Procedure of linking water with economic issues	Optimal allocations of water in a specific time period (e.g. study of the effects of pricing on water demand) or simulation of multi-period future scenarios (e.g., evaluation of the costs of future water allocation measures).
Assessment of cost recovery, including environmental and resource costs	Can be used to evaluate opportunity/resource costs of alternative water allocations between users.

Introduction

AQUATOOL (Andreu *et al.*, 1996) is a generic decision-support system (DSS) for integrated water resources planning and management, including conjunctive use of surface and groundwater. Originally, AQUATOOL was designed for the planning stage of decision-making with respect to water resource management in complex river basins. Subsequently, the system was expanded to incorporate modules relating to optimising utilities operation. With computer-assisted design modules through a graphical user interface, the complex water-resource system is represented as a network of objects[13] that can then be geographically referenced and linked to other GIS databases. Since AQUATOOL is both an optimisation and a simulation tool, it can be run to find optimal allocations in a specific time period or used for simulating multi-period future scenarios. AQUATOOL includes basin-management simulation and optimisation modules, an aquifer flow-modelling module, water quality simulator and two modules for risk assessment (Optirisk, Simrisk)[14].

[13]E.g. reservoirs, demands, aquifers, return flows, pumping stations, river nodes, surface water inflow, conduits/reaches, hydropower plants, artificial recharge, demand intakes. While objects in the graphical user interface vary, the approach is similar to modelling within MIKE Basin.

[14]http://www.upv.es/aquatool/

Interaction water economics

More recently, AQUATOOL has been expanded with a new economics module (Andreu *et al.,* 2005) that contains tools for systematically assessing the resource costs in a water resources system. The economics module in AQUATOOL is a post-processing module, i.e., is run based on simulations or optimisations of the hydrological system carried at the river basin scale. Together these tools can be used for economic analysis of water resource management as required by the WFD. For instance, opportunity cost of environmental measures, such as minimum flows in rivers or minimum volumes in reservoirs, can be calculated and assessed. Environmental and resource costs may be evaluated. The marginal resource opportunity costs (MROC) – resource costs – are defined as the cost for the system of having available one unit less of resource at a specific location and time. The marginal opportunity costs of environmental requirements (MOCER) – environmental costs – are defined as the cost for the system of increasing the environmental constraint in one unit.

The modelling tools for optimisation and simulation are used in conjunction. From optimising economic benefits (optimising approach), the model yields the optimal water allocation in the system at a particular time. On the other hand, from following a set of operating rules and institutional constraints (priorities and historical rights) the current modus operandi of the system (surface and groundwater resources, infrastructure, demands, etc.) may be simulated. With the gap between the economic value of the economically optimal water use (optimisation) and the current water allocation (simulation) any type of management strategy can be assessed.

In the stage of development and validation, AQUATOOL has been applied to the Segura and Tagus river basins in Spain. The River Basin Agencies in both study areas use a version of AQUATOOL for the efficient management of their water resources. The new module has been applied to the Jucar River Basin in Eastern Spain.

Scenarios and policy measures

The Decision Support System supports functionalities for both river basin characterisation, water utilities cost recovery, and cost-effectiveness analysis of programmes of measures required under WFD. The development of tools integrated within a DSS facilitates its application to different river basins, especially if a validated simulation model is available. Hydro-economic models provide a tool for a systematic assessment of two complementary measures of the resource opportunity cost and the opportunity cost of the management measures required to achieve the environmental objectives (eg., minimum streamflow) in a water resources system. Theses values, which change dynamically in space and time, can serve as indicators from which we could infer components of the resource and environmental costs required by the WFD.

References

Andreu, J., Capilla, J. and Sanchis, E. (1996). Generalised decision support system for water resources planning and management including conjunctive water use. *Journal of Hydrology* **177**, 269–291.

Andreu, J. *et al.* (2005). *DSS designed for integrated management of complex water resource systems including conjunctive use of surface and ground water. Presentation. Integrated River Basin Modelling and the EU Water Framework Directive.* International workshop, November 17–18, 2005, Vrije Universiteit Amsterdam, The Netherlands.

E. PIREN-Seine modelling toolbox – Ad hoc modelling chains for the implementation of the WFD

Table E.1 PIREN-Seine modelling toolbox

Kind of model (categorization)	Ad hoc linking of hydrological, chemical and algal biomass water quality models to mitigation cost calculations.
DPSIR description	Driver-Pressure (activity level nutrient loading), Pressure-State (nutrient loading – water quality)
Water use functions/emission sources included	Livestock, individual sewerage, rain water, municipal and industrial nutrient discharges.
Substances included	Nitrogen, Phosphorous
Comparison with emission and/or water quality standards	Yes. Risk index (1–3) based definition of good ecological status (current chemical quality, fish, diatoms, 2000–2015 baseline simulation, morphology, baseline evolution of discharges)
Priority substances of the WFD	No
Ease of use	Established models and modelling expertise required, as well as formatting of chained model input/output
Expected accuracy of results	Chained models subject to accumulation of uncertainty for each model link. Combined accuracy/joint uncertainty of combined models evaluated based on min/max assumptions (dry and wet seasons, loading with and without improvements in sewage treatment etc.)
Temporal scale and resolution of model and results (seasonality?)	Average annual water quality computed across models. Different models use different time steps; Avenant (annual), SENEQUE(10 days), PROSE (5 minutes), SIAM-1D (10 days)

(*continued*)

Table E.1 *Continued*

Spatial scale and resolution of model and results	Basin-wide scale with sub-catchment and river reach resolution
Down-scaling or up-scaling of results	Manual
Sensitivity and scenario analysis capabilities	Scenario analysis of baseline and supplementary mitigation measures
Uncertainty/simulation analysis capabilities	None
Economic demand and supply functions	None
Economic optimisation functionality (cost-effectiveness, efficiency analysis)	None. Cost-effectiveness analysis for ranking of measures by type of activity and sub-catchment.
Model transferability	Individual water quality models are basin specific. Chained modelling transferability depends on availability of calibrated models and model expertise for catchment in question.
Procedure of linking water with economic issues	Nutrient mitigation costs compared to changes in loadings and modular modelling of pressure-water quality state relationships for Seine water bodies (cost-effectiveness analysis)
Assessment of cost recovery, including environmental and resource costs	No

Introduction

An alternative approach to using fully integrated hydrological and economic models is to adapt the input and output formats of existing models in order to link them in "modelling chains". This has been the approach used by the Seine Normandy Water Agency in its PIREN-Seine programme. Driver-Pressure models of point sources of nutrient discharges to the Seine-Normandy basin has been linked to existing Pressure-State models of ecological status in rivers and estuaries. The advantage of using established models is coverage of both biological and physic-chemical parameters and the ability to simulate ecological status for the large number of sub-catchments that had already been calibrated in existing models. Models include:

SENEQUE: bio-chemical water quality model with a simple morphological and hydrological representation of surface waters, with coverage of almost the entire Seine-Normandy (as well as selected other basins). Predicts eutrophication (algal growth) based on two integrated sub-models of estuarine phytoplankton growth (Aquaphy) and degradation of organic material (HSB). GIS interface.

PROSE: also a bio-chemical water quality model comparable to SENEQUE but with increased functionality for various hydrological simulations, including finer temporal resolution. Implemented for the Seine River and its estuarine discharge.

SIAM-1D and SIAM-ELISE: bio-chemical water quality model for the Seine estuary (1D) and Seine bay marine environment (ELISE).

Avenant: Excel database/mass-balance model of nutrient loading and cost of nutrient mitigation works

Interaction water economics

The main focus of linking models to date has been to simulate scenarios of the risk of non-compliance with "good ecological status" in rivers and estuaries of Seine-Normandy given baseline/existing and supplementary measures. This is required in the economic characterisation of catchments under article 5 of the WFD. The chain of models is currently being used to conduct cost-effectiveness comparisons of basic and supplementary mitigation measures to comply with the WFD objectives. Calculation of mitigation costs and loading effect by type of source and sub-catchment location are coupled to the water quality models described above. Industry specific mitigation cost functions are used.

Scenarios and policy measures

Linked models provide a number of different outputs. (i) Spatially explicit distribution of algal biomass in water bodies in reference and supplementary measures scenarios. (ii) time series of predicted nutrient/biomass concentrations at any river node compared to ecological status classes (iii) "catchment profiles"for whole river reaches showing concentrations compared to ecological status classes (iv) years to "good ecological status" with and without supplementary measures at subcatchment level. Nutrient mitigation costs by source may be compared with any of these indices in order to conduct cost-effectiveness ranking of supplementary measures.

References

Blanc, S., Laurans, Y. and Grienche, G. (2004). *Elaboration d'un scenario tendanciel d'evolution de la qualité des cours d'eau du bassin de la Seine et des fleuves cotiers normands a l'horizon 2015*. Rapport final – novembre 2004. Agence de l'Eau Seine-Normandie, France.

Brignon, J-M. (2004). *Les modèles pression/impacts pour la Directive-Cadre Eau: Bilan des outils actuellement utilisés et des besoins futurs*. Rapport du Ministère de l'Écologie er du Développement Durable. Unité Modélisation et Analyse Economique pour la gestion des Risques (MECO), Direction des Risques Chronique.

Laurens, Y. (2005). *Developing ad-hoc modelling chains for the implementation of the WFD. Methodology and results on the Seine-Normandy basin. Integrated River Basin Modelling and the EU Water Framework Directive*. International workshop, November 17–18, 2005, Vrije Universiteit Amsterdam, The Netherlands.

F. Decision Support Tool for Irrigation (DSIRR)

Table F.1 DSIRR

Kind of model (categorization)	Single sector multiple-crop integrated hydro-economic model. Water quantity, but includes environmental impact/loading indices
DPSIR description	Driver (climate, policy scenarios)-Pressure (water demand, environmental pressure indices)
Water use functions/emission sources included	Irrigated agriculture water use; landscape and biodiversity (genetic diversity, soil cover), nutrients and pollutants(nitrogen balance, pesticide risk, energy balance)
Substances included	Pesticide index, nutrient index
Comparison with emission and/or water quality standards	Not directly, as emissions (nitrogen, pesticides) are index-based
Priority substances of the WFD	No
Ease of use	Graphical user interface for data input; Excel output tables and graphics. User does not need to programme/access GAMS code
Expected accuracy of results	Uncertainty analysis not supported so accuracy cannot be evaluated.
Temporal scale and resolution of model and results (seasonality?)	Depends on resolution of water supply, demand and farm production function data – usually seasonal or annual resolution.
Spatial scale and resolution of model and results	Data collected at river basin level. Water use data is not spatially explicit
Down-scaling or up-scaling of results	User defined aggregation. Any geographical scale is supported as water use data are not spatially explicit.
Sensitivity and scenario analysis capabilities	Scenario analysis and sensitivity analysis. Trade-off/ impact curves between multiple criteria supported (e.g. water pricing impacts on water use, farm income, seasonal farm employment, soil cover, nitrogen balance, pesticide risk etc.)
Uncertainty/simulation analysis capabilities	None
Economic demand and supply functions	Yes. Flexible definition of water supply function (block tariff structure), and multiple crop water demand functions

(*continued*)

Table F.1 *Continued*

Economic optimisation functionality (cost-effectiveness, efficiency analysis, cost-benefit analysis)	Mathematical programming optimisation in General Algebraic Modelling System (GAMS). Multi-criteria approach, optimising income, minimizing risk, labour input (etc.)
Model transferability	Generic model structure is transferable
Procedure of linking water with economic issues	Explicit modelling of farm production and multi-criteria objective function using GAMS
Assessment of cost recovery, including environmental and resource costs	Calculates income from water tariffs to water authorities. No environmental and resource costs

Introduction

The DSIRR "Decision Support for Irrigation" is a decision support (DS) for the economic-environmental assessment of agricultural activity focusing on irrigation. The program is currently used in the context of the EU WADI project (http://www.wadi.unifi.it/) to study the possible impact of WFD on irrigated agriculture in Italy. The model is billed as a useful tool for the development of basin level programmes of measures ("basin plans"). It has a fully integrated economic optimisation module for optimization of farm level objective function, which can include multiple farmer objectives (income, risk, farm labour, technical difficulty index). This multiple criteria analysis functionality makes it more suited for multi-stakeholder interaction than traditional optimisation models focusing only on maximization of farm profits. The model covers driver-pressure relationships of agricultural water use, using a series of indices to capture land-use and pollution pressures in addition to water use. The model does not cover water body states.

Interaction water economics

Water-crop yield functions on the demand side and water supply functions using flexible tariff structures are optimised for farm gross income and profit. Different mathematical programming techniques are used in tandem to optimise a multiple farm decision criteria (including minimization of risk, on-farm and hired labour, and a technical complexity index) subject a number of constraints on land, labour, capital, water balance and prices (Figure F.1). Resulting indexes for environmental and land-use pressures are also calculated (but not part of the objective function).

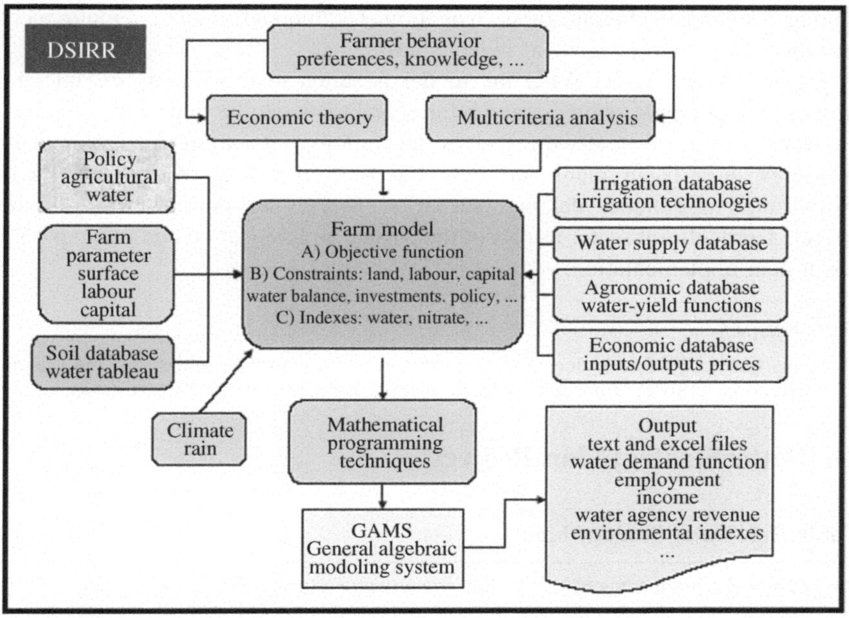

Figure F.1 DSIRR model structure. (Source: http://www.bo.ibimet.cnr.it/fileadmin/ibimet/repository/DSIRR_en.pdf)

While one of the model outputs is income generated from water tariffs there is no comparison with water supply costs, nor explicit treatment of cost recovery. The model is particularly suited to evaluating the impacts of water pricing policies on the farming sector, producing trade-off and sensitivity curves for prices against a number of non-economic objectives, including nutrient balance and pesticide use indices. As such it can be used to calculate the cost-effectiveness of water pricing as one instrument among many instruments and measures in a river basin programme of measures.

Scenarios and policy measures

DSIRR is not a generic overall system but a very specific tool to analyze water use and policy in agriculture with great attention to local situations in a macro context. Both short and long term scenario analysis are supported for a range of decision variables (farm enlargement/reduction, annuals crop mix); crop (plantation reduction/increase); irrigation (level change/techniques changes among existing techniques/new irrigation techniques); labour employment (seasonal/permanent); financial indebtedness.

In the context of the WFD the model is suited to to estimate ex ante the impact of water price policy in agriculture, and to assess the impact of subsidies, environmental taxes, water quotas, and technology innovation as supplmentary measures in a basin plan. As such, the tool may help Water and Basin Authorities to conduct detailed scenario analyses of policies towards the irrigated agricultural sector as part of the basin characterisation and participatory basin plan implementation.

References

Bazzani, G.M. (2005). A decision support for an integrated multi-scale analysis of irrigation: DSIRR. *Journal of Environmental Management* **77**(4): 301–314.

G. WaterStrategyMan Project

Table G.1 Water Strategy Man

Kind of model (categorization)	Multi-sector water allocation model with post-processing for economic optimisation. Includes a simple mass-balance water quality module
DPSIR description	Drivers-Pressures-State-Impact. See Figure F.1 for details
Water use functions/emission sources included	Types of water demand('user sites') supported: settlements, tourist, irrigation, industrial, animal breeding, hydropower, environmental, recreational and navigational requirements, drinking water, wastewater and desalinisation plants
Substances included	ammonia nitrogen, nitrate nitrogen, coliform bacteria, BOD, DO
Comparison with emission and/or water quality standards	No
Priority substances of the WFD	No
Ease of use	Non-commercial software. Windows based. Navigational panels, object manager. Map Editor for describing water network. Dedicated GIS map editor (not based on e.g. ESRI)
Expected accuracy of results	Based on input of normal year (average) monthly values. Not possible to enter full time series hydrological dataset. No calibration module.
Temporal scale and resolution of model and results (seasonality?)	Monthly

(*continued*)

Table G.1 *Continued*

Spatial scale and resolution of model and results	Catchment level. Reslution limited by the GIS data availability on water network. Size limits on network
Down-scaling or up-scaling of results	Limited. Whole catchment or individual water users
Sensitivity and scenario analysis capabilities	Scenario analysis module for evaluating packages of measures
Uncertainty/simulation analysis capabilities	No simulation based uncertainty analysis. Simulation capabilities for future years based on assumptions about growth rates in driver variables.
Economic demand and supply functions	Water availability defined by average normal year hydrological data. Water demand functions specified for each water user. Water demand functions defined as price elasticities specific to each water user.
Economic optimisation functionality (cost-effectiveness, efficiency analysis, cost-benefit analysis)	No optimisation. Economic post-processing module computes total benefit of water use, total environmental cost and rate of cost recovery.
Model transferability	Transferable generic modelling tool
Procedure of linking water with economic issues	Economic post-processing module
Assessment of cost recovery, including environmental and resource costs	Yes. Allows for the specification of environmental and resource costs based on French Agence de l'Eau approach of administrative charge rates per unit of concentration discharged

Introduction

The GIS-based Decision Support System developed for the WaterStrategyMan Project (WSM DSS) aims to assess the state of a water resource system in terms of sources, usage, water cycles (pathways) and environmental quality. In addition, it evaluates the effects of actions, on the basis of different scenarios, alternatives and policies. Water allocation is performed according to a set of demand and supply priorities reflecting the pricing system, social preferences, environmental constraints and development priorities. The WSM DSS can model water conditions in a given area and be used to estimate how much water is needed to meet the existing and projected demand, to determine what interventions are necessary, as well as when and where, and their cost. It can provide indicators of performance for selected actions under potential availability and demand scenarios, and use them to rank the scenarios. The DSS supports the user to assess the functionality and performance of the water system within the entire region of application as well as individual points of interest (ProGEA, 2004).

Figure G.1 gives an overview of the decision-support system structure which is based on the driver-pressure-state-impact-(response) framework adopted by the European Environmental Agency.

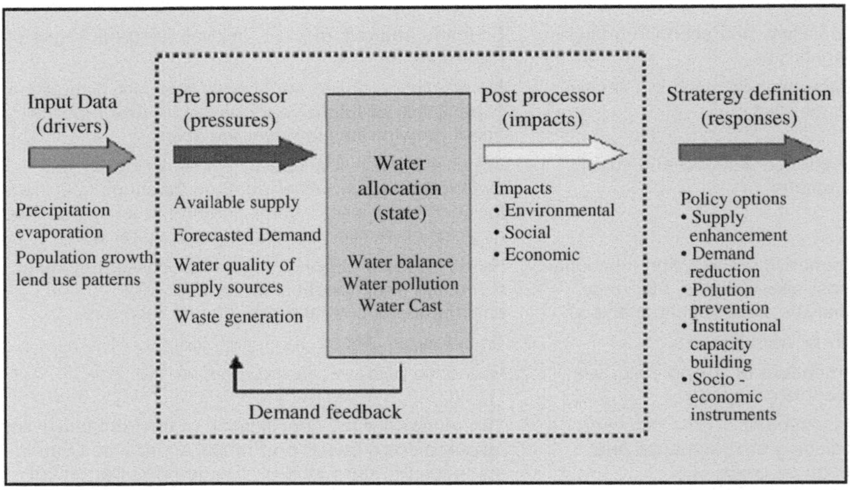

Figure G.1 Water StrategyMan DSS structure. (Source: http://environ.chemeng.ntua.gr/wsm/Default.aspx?t=1)

Interaction water economics

The DSS aims to capture all the components of full costs of water, including supply costs, resource costs and environmental costs. Marginal value of water quantity can be directly defined by the user or estimated according to the specific water user as marginal cost of the most expensive source under use (urban uses), it can depend on crop market price, average annual yield, annual water supplied or the alternative value of land. The estimation of environmental costs is based on the French Agence de l'Eau approach using administratively specified "charges" per unit of concentration. Direct supply costs are defined as operating costs, maintenance costs, capital costs, administrative costs and other direct costs.

Scenarios and policy measures

The evaluation module of the DSS allows the user to specify a number of different management options alone or in concert:

- *Supply enhancement options*, intended to increase available water quantities during drought; they concern structural interventions which attempt to enhance water supply;
- *Demand management options*, aiming at decreasing water demands through various conservation techniques and use limitation;
- *Socio-economic measures* needed to mitigate impacts, also by means of socio-economic instruments, such as pricing and changes in the regional development priorities.

For each simulation a series of management indicators are calculated within three different categories. *Environment and Resources:* Minimum flow requirement coverage, Desalination and Reuse Supply Share, Dependence on Imported Water, Groundwater Exploitation Index and Non Sustainable Groundwater Production Index.

Efficiency: The Efficiency group is characterised by the rate of demand coverage for each kind of water use in the region, from domestic to hydroelectrical.

Economics: Total Benefit from Water Use, Total Environmental Cost and Rate of Cost Recovery.

Of the hydro-economic models reviewed this is the only one designed to meet the specific WFD requirements. One drawback is that the system requires quite a lot of pre-processing of data in order to arrive at average values of coefficients required by the model computations.

References

ProGEA (2004). The WaterStrategyMan DSS. A Comprehensive Decision Support System for the Development of Sustainable Water Management Strategies http://environ.chemeng.ntua.gr/wsm/Uploads/Deliverables/ThirdYear/Deliverable_21_4.pdf.

IWA Publishing's authorised EU representative for General Product Safety Regulations is Diane D'Arras, 15 rue Duret, 75116 Paris, France, e-mail: safety@iwap.co.uk.

Printed and bound by CPI Group (UK) Ltd, Croydon, CR0 4YY
12/05/2026
02108866-0004